Negotiated Learning

*Collaborative Monitoring in
Forest Resource Management*

Edited by
Irene Guijt

RESOURCES FOR THE FUTURE
Washington, DC, USA

Printed in the United States of America

An RFF Press book
Published by Resources for the Future
1616 P Street NW
Washington, DC 20036–1400
USA
www.rffpress.org

Library of Congress Cataloging-in-Publication Data

Negotiated learning : collaborative monitoring in forest resource management / edited by Irene Guijt.
— 1st ed.
    p. cm.
Includes bibliographical references and index.
ISBN 978-1-933115-38-2 (pbk. : alk. paper)
ISBN 978-1-933115-39-9 (hbk : alk. paper)
1. Forest monitoring—Case studies.   2. Forest management—Case studies.   I. Guijt, Irene.
SD391.N44 2007
634.9'2--dc22   2006030028

The paper in this book meets the guidelines for permanence and durability of the Committee on Production Guidelines for Book Longevity of the Council on Library Resources. This book was typeset by Integrated Book Technology. It was copyedited by Sally Atwater. The cover was designed by Henry Rosenbohm. Cover photo by Yayan Indriatmoko.

ISBN 978-1-933115-38-2 (paper)   ISBN 978-1-933115-39-9 (cloth)

About Resources for the Future *and* RFF Press

RESOURCES FOR THE FUTURE (RFF) improves environmental and natural resource policymaking worldwide through independent social science research of the highest caliber. Founded in 1952, RFF pioneered the application of economics as a tool for developing more effective policy about the use and conservation of natural resources. Its scholars continue to employ social science methods to analyze critical issues concerning pollution control, energy policy, land and water use, hazardous waste, climate change, biodiversity, and the environmental challenges of developing countries.

RFF Press supports the mission of RFF by publishing book-length works that present a broad range of approaches to the study of natural resources and the environment. Its authors and editors include RFF staff, researchers from the larger academic and policy communities, and journalists. Audiences for publications by RFF Press include all of the participants in the policymaking process—scholars, the media, advocacy groups, NGOs, professionals in business and government, and the public.

About the Center for International Forestry Research

THE CENTER FOR INTERNATIONAL FORESTRY RESEARCH (CIFOR) is a leading international forestry research organization established in 1993 in response to global concerns about the social, environmental, and economic consequences of forest loss and degradation. CIFOR is dedicated to developing policies and technologies for sustainable use and management of forests, and for enhancing the well-being of people in developing countries who rely on tropical forests. CIFOR is one of the 15 Centers of the Consultative Group on International Agricultural Research (CGIAR). With headquarters in Bogor, Indonesia, CIFOR has offices in Brazil, Burkina Faso, Cameroon and Zimbabwe, and it works in over 30 other countries around the world.

Contents

Part IV. Dealing with Difference

Part V. Adapting Monitoring Processes

Part VI. Conclusions

Foreword

*T*HIS IMPORTANT BOOK BREAKS NEW GROUND. Monitoring has often been associated with upwards accountability; the measurement of centrally determined indicators; regular, repetitive, and routine reporting; and, quite often, accumulated sediments of information that are unread and unused. In English, the word "monitoring" already has unfortunate undertones of surveillance. As noted by the contributors to this book, it can also translate into other languages as words that imply control and even policing.

For CIFOR (the Center for International Forestry Research), which supported the research initiatives discussed in this book, monitoring was structured around a hierarchy of four concepts: principles, criteria, indicators, and verifiers. Attempts were made to apply these by teams of facilitators in a participatory action research mode in Bolivia, Brazil, Cameroon, Ghana, Indonesia, Malawi, Nepal, the Philippines, and Zimbabwe. Their experiences have generated rich, reflective, and insightful accounts of learning and change that show us new potential.

Though the case study experiences naturally differ, their commonalities are significant. First, teams found that detailed use of indicators, while logical from a research and information perspective and prevalent in the world of monitoring and evaluation, impeded learning and contributed little to resource management. Of the four theoretical concepts that shape CIFOR's indicator framework, it was only "verifiers," close to the realities of local participants, that made some sense. Second, the teams' participatory approaches had to evolve to be relevant in complex, diverse, and dynamic local conditions. The teams had to not only facilitate but to negotiate and improvise, adopting a pluralism of approaches rather than applying preset methodologies that are so prevalent in the world of monitoring and evaluation. Third, the teams found themselves co-learners.

Paradigmatically, this is a new world for monitoring. It goes beyond the prevalent monitoring of participatory monitoring and evaluation, in which local people participate within frames that are externally determined, observing and measuring indicators that are passed down to them, and then passing the findings back up. Instead, this book redefines monitoring as a process of "socially negotiated learning," which requires information seeking, critical analysis, and process facilitation in equal measure. Monitoring focuses on co-learning and adaptive management by local actors. Local actors include not just resource users but others like forest rangers and government officials who influence resource decisions and implement resource use legislation.

Monitoring in this mode fits, flows from, and functions in the world of local complexity, diversity, and dynamism. Unpredictability is inherent; surprise is expected and sought. Rigor is not in methodological rigidity but in relevance of learning. This in turn derives from the quality of process, interactions, and relationships. The caliber, orientation, behavior, and attitudes of facilitators, and the organizational understanding and support they receive, are crucial.

Negotiated Learning: Collaborative Monitoring in Resource Management is about much more than participatory *forest* management. Its lessons apply to all participatory *natural resource* management. Beyond that, both deeper and broader, it has messages for all development professionals about how "we" think of and go about development. The experiences it describes challenge central mindsets, the paradigm of preset research, and the rationality of bureaucratic accountabilities. Those currently frustrated by ineffective ways of working may now see collaborative monitoring and negotiated learning as personal, professional, and institutional paths to innovation with procedures, accountabilities, and priorities.

This volume should lead to decisive and imaginative action not only in adaptive collaborative management but in development more generally, as it opens up and illuminates the world of potential through facilitation, process, relationships, and social learning. May it inspire many others to follow where it leads.

ROBERT CHAMBERS
Research Associate,
Institute of Development Studies,
University of Sussex, United Kingdom

Acknowledgments

R AVI PRABHU OF CIFOR WAS PARTICULARLY supportive in championing the idea of this collection of experiences and providing critical comments on the draft. I am grateful to Carol Colfer for advice and encouragement during the completion stage. I am indebted to Marlene Buchy of the Institute of Social Studies (the Netherlands), Wouter Leen Hijweege of Wageningen International (the Netherlands), and Dr. Julian Gonsalves (the Philippines) for their insightful comments on draft versions of this volume, in particular the introductory and summary chapters. An anonymous reviewer helped improve the structure and sharpen the style. Atie Puntodewo and Gideon Suharyanto of CIFOR, Bogor, kindly helped by providing all the figures. Don Reisman, Grace Hill, and Miriam Dowd of Resources for the Future processed the manuscript with care and commitment; my thanks. Sally Atwater picked out the last mistakes with her editing. Any errors remain solely my responsibility.

The funding support of the Asian Development Bank, the Department for International Development (United Kingdom), the European Union, the U.S. Agency for International Development, and CIFOR made it possible to undertake the fieldwork on which these experiences are based and then to analyze them and commit them to paper. This book contributes to the IUFRO Task Force "Improving the Lives of People in Forests," which seeks to further the issues discussed by the authors.

Contributors

SAMUEL ASSEMBE is a jurist and research assistant with CIFOR's Forest and Governance team in Cameroon.

GUILHERMINA CAYRES is an agriculturalist who works as a consultant on socio-environmental development of smallholder communities in the Amazon, focusing on rural development, participatory approaches, gender issues, and criteria and indicators for assessing sustainability.

PETER CRONKLETON is an American anthropologist who coordinates community forestry research in Bolivia specifically and in the western Amazon generally. He also works with grassroots networking and social movements in Latin America.

SUSHMA DANGOL is a Nepali forester and researcher. She worked with NORMs during the research highlighted in this book and is currently working with NewERA, both Nepali NGOs and partners in CIFOR's Adaptive Collaborative Management research project.

MARITEUW CHIMÈRE DIAW is a Senegalese anthropologist working for CIFOR and leading the Forest and Governance team in Central Africa.

The late **SAMUEL EFOUA** was a research technician with the Forest and Governance team in CIFOR Cameroon until his death in 2005.

KRISTEN EVANS has worked with forest communities in Bolivia for several years, first as a Peace Corps volunteer and then as a research consultant. She is researching participatory methods in Bolivia and Vietnam with CIFOR.

IRENE GUIJT is a Dutch independent consultant specializing in organizational learning and sustainable rural development. She was on the International Scientific Steering Group of CIFOR's Adaptive Collaborative Management research project from 2000 to 2002.

HERLINA HARTANTO is an Indonesian scientist who has been working on community forestry issues in Indonesia and Philippines for the past eight years. She was the Philippine country coordinator for CIFOR's Adaptive Collaborative Management project.

JUDITH KAMOTO coordinated the Adaptive Collaborative Management project in Malawi. She is a lecturer in Forest Extension and Rural Development at the University of Malawi, Bunda College of Agriculture, Malawi.

ROBERT E. KEATING III spent three years working with community-based forest management plans in Bolivia. He travels to Africa, Asia, and Latin America as a financial analyst and field accountant for Chemonics International Inc.

CHIRANJEEWEE KHADKA is a social forester and researcher from Nepal. He worked with the Nepali NGO NORMs during the CIFOR-led research highlighted in this book.

WITNESS KOZANAYI is a consultant with the Adaptive Collaborative Management Zimbabwe team. He has an agricultural background and experience in rural livelihoods and participatory methodologies.

TRIKURNIANTI (YANTI) KUSUMANTO is an agriculturalist who has been coordinating CIFOR's Adaptive Collaborative Management field research in Jambi, Sumatra, since 2000.

WILLIAM MALA is an agroecologist working with CIFOR's Forest and Governance team in Cameroon.

CYNTHIA MCDOUGALL is a senior associate with CIFOR and has been the project leader of CIFOR's Adaptive Collaborative Management research project in Nepal since 1999.

TENDAYI MUTIMUKURU is a Zimbabwean specialist in social learning. She worked as a participatory action researcher with CIFOR in Mafungautsi State Forest.

WESTPHALEN NUNES has worked on participatory and collaborative approaches with several government agencies and NGOs since 1992. He is technical director of the Assessoria Comunitária e Ambiental, a consultancy located in Belém, Brazil, which supports communities in environmentally based development.

RICHARD NYIRENDA is a Zimbabwean forester who worked as a member of the Zimbabwe Adaptive Collaborative Management team. He works in the United Kingdom, where he is coordinating the activities of a Zimbabwean charity, the Dzimbahwe-Umuzi Trust.

HEMANT OJHA is a founder of ForestAction, Nepal, and is an active member of the Adaptive Collaborative Management team in Nepal.

PHIL RENÉ OYONO is a Cameroonian sociologist and a member of CIFOR's Forest and Governance team in Cameroon.

KRISHNA P. PAUDEL is a founder of ForestAction, Nepal, as well as an active member of the Adaptive Collaborative Management team in Nepal.

BENNO POKORNY worked in the regional CIFOR office in Belém, Brazil from 1998 to 2003, researching opportunities for sustainable forest management in the Amazon. Still associated with CIFOR, he is a member of the Faculty of Forest and Environmental Sciences at the University of Freiburg, Germany, where he coordinates international research about community forestry and rural extension, mainly in South America.

MAGNA CUNHA DOS SANTOS is a Brazilian community forester and grassroots advocate working for CTA (Centro dos Trabalhadores da Amazônia), the Amazonian Workers Center, a Brazilian NGO in Acre.

MARIANNE SCHMINK is an American anthropologist and professor at the University of Florida; she has linked CIFOR scientists with the Adaptive Collaborative Management field team in Acre, Brazil.

SAMANTHA STONE is a French American anthropologist who conducted her doctoral research in the Acre Adaptive Collaborative Management site. She works in Australia as a researcher with CSIRO in Townsville.

PART I

Introduction

CHAPTER 1

Strengthening Learning in Adaptive Collaborative Management
The Potential of Monitoring

Irene Guijt

NATURAL RESOURCE MANAGEMENT THAT IS LINEARLY planned, externally controlled, and focused on a single interest is widely considered ineffective. Much has been said and written about the dynamics of resource management and the need for participatory management. This extends to the specific case of forests, which are home to many users, supply multiple products, and serve diverse functions—all of which need to be considered. Forest management that is both adaptive and collaborative is therefore appealing and actively pursued. On the ground, decision making and implementation have become more participatory and multifunctional, and inflexible implementation has given way to more dynamic management (Jiggins and Röling 2000; Lee 1999; Dovers and Mobbs 1997; Gunderson et al. 1995; Borrini-Feyerabend et al. 2000; Ghimire and Pimbert 1997; Hinchcliffe et al. 1999; Defoer et al. 1998; Roe et al. 1999; Colfer 2005a; Leach et al. 2005).

However, practice in other areas is slower to change. One issue that has received very little attention to date is how to ensure that adaptation in resource management comes from learning—and is not just a result of crisis or unilateral decisions. This means asking and answering difficult questions. Who needs to learn what for resource management to improve? How can such learning be created and sustained? Where will decisions be made? What signals can forest users observe to know what changes are needed?

Thus far, a common response has been to invest in data collection, with the creation of vast databases and information storage systems. This embodies a knowledge management perspective rather than an organizational learning perspective (Easterby Smith and Lyles 2003). Rarely, however, do the resulting data sets fuel the type of timely and negotiated learning that is required. Adaptation often ends up as ad hoc, incidental changes based on unclear evidence and opaque decision making. Forests and forest communities lose out as a result.

This book seeks to expand our understanding about how monitoring can create meaningful learning for adaptive management. The contributors reflect critically on their practical experiences as part of an adaptive and collaborative approach to forestry management. Their efforts took place within the context of a global research program of the Center for International Forestry Research (CIFOR). The action research led to a wide range of local impacts, analyzed in detail by Colfer (2005a[1]). This book goes one step further in discussing the conscious learning processes that are needed to make such impacts possible.[2]

It is the first volume that brings together analyses of hands-on attempts at monitoring as something beyond data collection. Importantly, these case studies come from longer experiences and thus reflect the difficulties that emerge after initial enthusiasm wanes. The contributors' collective message is that data can form the basis for improvements only if those involved invest in the social processes of analysis and decision making and in the technical and political capacities required to make such processes possible.

The 11 experiences in this book should interest all those who are engaged in adaptive collaborative management and forest management in general, particularly in the Southern Hemisphere. The in-depth descriptions can inspire community development professionals and facilitators, as well as those in development agencies that fund sustainable natural resource projects. The frank narratives offer insights to all those involved in organizational learning and evaluation, be it within funding agencies or on the ground as researchers, facilitators, or active citizens. The concluding chapter, which summarizes lessons from the experiences, may also prove useful for impact assessment practitioners.

This chapter introduces the practical experiences that follow. First, the concept of adaptive collaborative management (ACM) as conceptualized by CIFOR is explained. This paves the way for an explication of how learning and monitoring are embedded within ACM. The next section describes the generic approach of CIFOR's ACM research teams in creating collaborative monitoring. The approach involves a combination of constructing the information base and creating opportunities for analysis—as well as facilitating both these processes. The chapter ends by introducing the diversity of practice that the subsequent chapters illustrate.

Adaptive Collaborative Management[3]

The central concern behind the case studies was how to create the conditions necessary to sustain forests and forest communities around the globe.[4] In 1999, CIFOR researchers launched a research program to gain insight into the conditions, processes, and institutional arrangements for collaboration and conscious social learning in community forest management and related human and forest outcomes. They labeled the type of forest management they were seeking to create "adaptive and collaborative." The first research phase took place from 1999 to 2002 and involved 30 researchers in 9 countries (Table 1-1). Followup work has since extended the research into 11 countries (Figure 1-1).

Table 1-1. *Initial ACM Participatory Action Research Sites*

Countries	Project	Funding agency	Years	Followup
Nepal, the Philippines, Indonesia	Planning for the sustainability of forests through adaptive comanagement	ADB (RETA 5812)	1999–2002	Nepal (IDRC)
Zimbabwe, Malawi	Developing indicators-based collaborative monitoring arrangements to promote adaptive community based forest management	DFID	1999–2002	DFID
Zimbabwe, Malawi, Ghana, Cameroon	Developing collaborative monitoring for adaptive comanagement of tropical African forests	European Union (B7-6201/ 99-05/FOR)	2000–2003	Zimbabwe (CIDA), proposal with EU for other countries
Brazil, Bolivia	Compatibility of C&I-based instruments for certification, monitoring, and auditing the sustainability of forest resource use; management of secondary forests by smallholders in northeastern Pará; Bolivia Adaptive Collaborative Management Program	GTZ, PPG-7 (ProManejo), USAID, CIFOR	2000–2003	Two proposals under discussion with EU and German government

Notes: ADB Asian Development Bank; CIDA Cultural Industries Development Agency (U.K.); DFID Department for International Development (U.K.); GTZ Gesellschaft fur Technische Zusammenarbeit; IDRC International Development Research Centre; USAID U.S. Agency for International Development.

Forest-based crises in all sites required an approach that allowed the simultaneous development and application of ideas based on existing community forestry arrangements. However, such arrangements usually involve only local communities (often superficially) and exclude many other stakeholders—forest rangers, policymakers, logging companies, and so forth—whose participation is critical for sustained success. Both the urgency and the need for more inclusiveness led to CIFOR's choice of participatory action research (PAR) as the methodological vehicle (discussed below in more detail).

The choice of research approach that builds on in-situ conditions and relationships spawned great diversity among the resource management processes that emerged. Nevertheless, they had several starting points in common. Paramount among these was the recognition of complexity and diversity of forest and human systems as givens. This meant finding a way to be open to and deal with surprises, since "complex systems, because of internal interactions and feedback mechanisms, tend to generate 'surprises'" (Ruitenbeek and Cartier 2001). A second common element was a concern for equity (cf. Colfer 2005b). Adaptive collaborative management would work only if local communities (and their marginalized

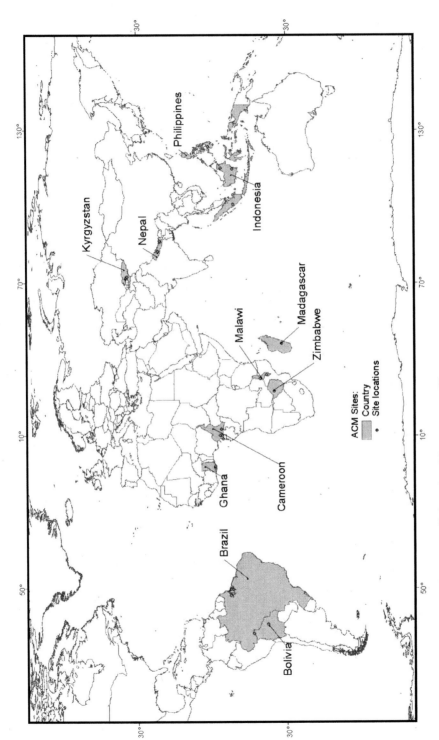

Figure 1-1. *CIFOR's Adaptive Collaborative Management Sites*

subgroups) had a place at the negotiation table and if the uneven playing field was leveled. This meant creating effective communications between individuals and groups in diverse and dynamic power relationships.

The third common element of the ACM research processes takes center stage in this book: the need for continual feedback. Explanations for the all-too-common policy failures were hard to find, and as Colfer states, "Critical opportunities to learn from failures were missed, and malfunctioning management systems continued to hobble along, either ineffective or damaging" (2005a, 3). CIFOR initially focused on (simplified) adaptations of earlier work on criteria and indicators (Ritchie et al. 2000), later extending this to include other monitoring approaches able to catalyze the kind of social learning that would lead to the adaptations that were so critical.

These features led CIFOR to define adaptive collaborative management as

> a quality-adding approach whereby the people or groups who use, control or in some way have interests in a forest, agree through a process of participatory action research to act together when they draw up plans for their forests. These plans are then implemented with awareness that they may not fulfill their stated objectives. In this process, it is important for people to observe and learn from implementation, together as groups, as improvements in the plan are negotiated and alternatives are sought. (Prabhu et al. 2002)

Table 1-2 summarizes the types of process changes that were anticipated as the action research teams progressed. Information, communication, and learning—three ideas central to this book—figured prominently in the initial understanding of ACM. All research efforts were based on the assumption that continual improvements were possible if resource users and managers made conscious efforts to undertake joint learning, be it about the impacts of past actions, uncertainties, or new phenomena. If, in this manner, resource management built on local knowledge and expertise and became self-improving, then ACM might have true potential to improve human and forest well-being simultaneously.

However, developing such monitoring systems is far from easy. It is also a relatively young area of theory and practice.[5] This brings us to the main question that this book addresses: What is needed to ensure that monitoring leads to learning and in turn can lead to the ongoing improvements sought through ACM?

Making Adaptation Conscious

Adaptive collaborative management is intended to be flexible—improving resource management practices, incorporating new partners, adjusting the roles of existing partners, modifying decision making, refocusing priorities and strategies, and so forth. But what should trigger these adjustments and indicate what shape the changes must take?

Changes often result from crises, through power plays, or as dictated externally by new laws or funding conditions. But they can also occur through reflection on

Table 1-2. *Shifting from Low to High Adaptiveness and Collaboration*

	Low	High
Adaptiveness	• Few information flows, blockages, imbalances	• Effective information flow among stakeholders
	• Learning is ad hoc	• Intentional learning
	• Resistance to learning and change	• Institutional willingness or attitude and capacity to learn and respond to learning
Collaboration	• No stakeholder meetings	• Joint decision making
	• Decisions made by select few without consultation	• High diversity among participants
	• Limited diversity among participants	• High level of mutual respect and trust
	• Distrust or lack of respect	• Cooperation in management tasks
	• High levels of conflict, perhaps violence and destruction	• No or manageable levels of conflict

Source: McDougall 2000.

information and through collective "sense making." Such deliberative processes can be structured, through long-term data sets that reveal trends. It can be less structured, through unexpected occurrences, the surprises and discomfort that alert people to a discrepancy between what they thought would happen and what is, in fact, the case. Either way, it is a social process.

Learning, Defined

Learning within ACM is defined in this book as a social process by which communities, stakeholder groups, or societies learn how to innovate and adapt in response to changing social and environmental conditions (Woodhill 2004; also see Box 1-1). This means that it is not just placing one's faith in scientific discovery or hoping that market mechanisms will iron out imbalances. Rather, it is an interactive perspective that "brings people into negotiation with each other over their values, their goals, their differing interests and the development of collective interests and common strategies for action" (Woodhill forthcoming). ACM links to other discourses that focus on the importance of societal-level change and a perspective of social learning as seeking to alter underlying structures and norms (Milbrath 1989; Waddell 2005; Leeuwis and Pyburn 2002; Woodhill 2002). Central in these is the idea of critical intent, which refers to an individual disposition in a process of inquiry but also to a social consciousness of the individual (Grundy 1982).

Two specific forms of learning are important in ACM: retrospective learning and anticipatory learning. Retrospective learning focuses on monitoring past actions, each experience offering lessons to shape the next improved iteration of resource management. Anticipatory learning allows one to prepare for change by anticipating the range of changes that might occur, their consequences, and

---

Box 1-1. From Learning to Social Learning

"A society that is unable to innovate in response to a changing environment runs the risk of crisis, if not annihilation. History does not lack examples of societies that have met this fate. Any social change requires learning of some form, but the question here is how societal-wide learning processes can be more, rather than less, effective and how this can be facilitated. Social learning seeks an alternative to two classical strategies for governance: (1) that government and experts should make decisions for society and 'solve our problems' or (2) believing that social change should be left largely to market forces with minimal guidance by government. Failure at both ends of this spectrum of governance mechanisms has fed the interest in social learning and more participatory forms of democratic governance."

Source: Woodhill 2004.

---

available options for dealing with any problems. Scenario-based techniques are the main tool for anticipatory learning. Although anticipatory learning is important for dealing with the complex and risky nature of ACM, this volume focuses on retrospective learning through collaborative monitoring.

Monitoring for Learning

Monitoring itself goes by many definitions, most of which stress systematic rather than ad hoc or one-off processes like evaluation or impact assessment. In monitoring, the learning is institutionalized; it becomes the norm. The term is also generally associated with data collection to satisfy information needs. However, understandings differ about whether monitoring includes analysis. This leads to varying views about whether monitoring includes assessing merit or value, and therefore how it relates to decision making.

Since the mid-1990s, ideas about making monitoring more participatory or collaborative have taken root (Abbot and Guijt 1998; Estrella et al 2000; Estrella and Gaventa 1997). Participatory monitoring recognizes the central role that local people can play in planning and managing their use of the environment. It reflects a logical evolution of the participatory approaches to resource appraisal that have developed over the past two decades (cf. Chambers and Guijt 1995). Participatory monitoring shifts the emphasis away from externally defined and driven programs and stresses the importance of a locally relevant process for gathering, analyzing, and using the information. It means involving (groups of) people in aspects of monitoring in which they have not previously been involved and creating conditions so that they can dictate the focus, means, and rhythm of the learning that occurs.

Much of the literature on collaborative monitoring is simplistic about the process and unrealistic about its potential. Compare the two statements in Box 1-2. The first statement represents a commonly held, optimistic view based on the image of participatory monitoring and evaluation as a fairly foolproof solution to the problems caused by linear planning, imposed solutions, and marginalized stakeholders. The second statement reflects the views held by a small but growing group of people who have worked with collaborative monitoring. By tracking and analyzing changes, relations, and problems, they show how forest stakeholders can construct new strategies and management options. But they clearly say that this requires time, capacity, and commitment. Thus, while not detracting from the possible benefits that collaborative monitoring can offer, these practitioners add a solid dose of realism. Above all, they show that learning matters in natural resource management and has implications for monitoring approaches.

Monitoring that involves critical reflection on information—and not just data collection—is pivotal. Continual information input—about the state of the resources, about how they are being used, about how to work together in making decisions, and so much more—is crucial. But if collective action is to ensue, then collective sense making through critical analysis of information is essential.

Just as joint forest planning has proven to be complex (cf. Colfer 2005a), so has the practice of collaborative monitoring. Some questions are of a practical, how-to nature. How do we select what to monitor when there are such diverse information needs? How can we analyze qualitative and quantitative information? How do we decide who should be involved and what their roles should be? How do we deal with deeply embedded social inequities when trying to bring together different social groups? What can we do to ensure that the monitoring system remains relevant and makes a difference?

But more fundamental questions also need answers. For example, one recurring dilemma is whether accountability-focused information can be combined with a strategic learning orientation. How can an organization, an alliance, a network, or a project strike an optimal balance between accountability and learning (Edwards 1997; Lindenberg and Bryant 2001; Pettit and Roper 2003; Wallace and Chapman 2003)? Also important is the question of how to engender a renewed appreciation of dialogue and conversation that enable learning as integral to monitoring. Perhaps most important, if empowerment is a concern, is how to

---

Box 1-2. Simple or Grounded? Different Views on Collaborative Monitoring

1. " ... participatory monitoring and assessment ... will further enhance the involvement of affected communities and stakeholders in ... projects, provide for better correction during implementation, and ensure that lessons are articulated and learned by the participants themselves. The involvement of key stakeholders contributes to community 'ownership,' helps build consensus about the project's approach, and promotes mutual understanding." (UNDP, undated)

2. "Collaborative monitoring has risks and limitations that communities and all other partners must understand. As a mechanism to enhance understanding and decision making, it has great potential. However, if there is no time for communities to acquire the necessary skills and insights, then it loses its local value and becomes yet another externally driven evaluation tool" (Santos et al., Chapter 3).

make participation in monitoring a process in which powerful social groups do *not* dominate and dictate the conversations in which analysis takes place and from which decisions emerge.

Essential Activities for Collaborative Monitoring

Given the above considerations, collaborative monitoring for ACM is defined in this book as a process of conscious information seeking followed by shared critical analysis to inform collective decisions that affect resource management.

On-the-ground conditions and needs determine variations in design and implementation. In the experiences discussed here, some authors identified information needs using what is known as the criteria and indicators framework (Chapters 2, 3, 7, and 10). However, other approaches are possible (Chapters 5 and 6). For shaping the processes of sense making and communication of insights, the authors were all inspired by the action-learning cycle of participatory action research. Both these processes required facilitation, a role that the authors took up and often shared with other stakeholders. Facilitation was critical, given the multiple interests at play in all the case studies.

When linked in an interactive cycle, these three[6] activities—information seeking, sense making, and process facilitation—were the indispensable building blocks of collaborative monitoring. The diversity in monitoring practice is shaped by how the stakeholders chose to answer the four questions in Figure 1-2, based on their historical and institutional context. This diversity is discussed in more detail at the end of this chapter. First, the three building blocks are described in terms of the research teams' initial perspectives.

The Information Base: Criteria and Indicators

From 1994 to 2000, CIFOR developed and tested criteria and indicators (C&I) for assessing the sustainability of community-managed forests[7] (Ritchie et al. 2000; Purnomo et al. 2000; CIFOR 1999a, 1999b; Colfer et al. 1999a, 1999b; Mendoza et al. 1999; Prabhu et al. 1999). The research results were described as "a participatory tool for sustainable community managed forests"[8] (Ritchie et al. 2000, 1). The C&I framework was intended for use by community-based forest managers and/or practitioners and partners. The C&I that are produced at the local level ideally provide a framework for monitoring and assessing important changes, thereby feeding information and learning back into the community-managed forest system and guiding future action toward sustainability. As conceived by CIFOR, the framework would help organize local and scientific knowledge into an effective forest and forest management "health check."

The framework is structured around four concepts organized in a hierarchy: principles, criteria, indicators, and verifiers[9] (Table 1-3). This hierarchy is supposed to be constructed by the stakeholders who will use the resulting information. Various methods to help construct a C&I set are suggested in the

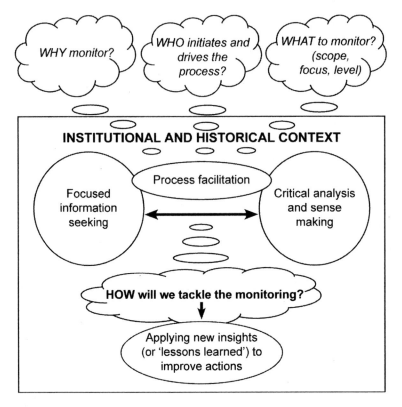

Figure 1-2. *Building Blocks and Core Questions that Shape Collaborative Monitoring Activities*

chapters that follow, such as visioning, participatory mapping, and card sorting (Ritchie et al. 2000).

The principles lay the basis for defining the criteria, which in turn determine the indicators and related verifiers. While serving as the starting point, the principles also form the final point of integration of information analysis. In CIFOR's original framework, four principles were listed as essential for sustainable community forestry management: ensuring community well-being, people's well-being, forest landscape health, and a supportive external environment. However, principles would be based on people's visions of their forests and thus vary by site.

In this volume, several authors critically discuss their attempts to work with the C&I framework at the community level. They describe the uneasy and uncomfortable fit of the overly detailed and comprehensive hierarchy with the collaborative and more open-ended processes of ACM (Chapters 2 and 3). If an ACM initiative starts small, based on a specific issue, then a comprehensive approach to data identification may not be necessary (Chapters 5, 6 and 10). Where it concerns new forest management initiatives, such as in Brazil (Chapter 3) or Cameroon (Chapter 9), then the C&I framework may be more worthwhile considering.

Table 1-3. *Defining Criteria and Indicators*

Concept	Definition	Examples
Principle	A fundamental truth or law as the basis of reasoning or action. In the context of sustainable forest management, principles provide the primary framework for managing forests in a sustainable fashion.	"Ecosystem integrity is maintained or enhanced." "Human well-being is ensured."
Criterion	A standard by which something is judged. A criterion can therefore be seen as a second-order principle, one that adds meaning to a principle and makes it operational, without itself being a direct measure of performance.	"Processes that maintain biodiversity are maintained."
Indicator	Any variable or component of the forest ecosystem or the relevant management systems used to infer attributes of the sustainability of the resource and its utilization. Indicators should convey a "single meaningful message" or information.	"Landscape pattern is maintained."
Verifier	Data or information that makes indicator identification easy. Verifiers provide specific details that reflect a desired condition of an indicator, thus adding meaning, precision, and usually also site specificity to an indicator.	"Areal extent of each vegetation type in the intervention area relative to area of the vegetation type in the management unit."

Source: Ritchie et al. 2000.

Whether or not the context requires a more comprehensive approach, a significant obstacle appears to lie in the four-layer structure of CIFOR's C&I approach, which conceptualizes information needs to a level of detail that most stakeholders do not seem to need or desire (Chapters 2 and 3). Other problems with the framework parallel the limitations of any indicator-focused approach. These include the inability to explain causes behind observed changes and a tendency toward overly extensive indicator sets that consume time and resources (with inevitable loss of data quality). Furthermore, the short shelf life of some information needs sits uneasily with the requirement of trend analysis to monitor the same set over a sustained period. Finally, indicators can deal only with anticipated phenomena of change (Davies 1998) and are weak in exactly the area where ACM requires input—the surprises.

Although CIFOR's methodology stresses the importance of participatory identification of this informational hierarchy, the existing documentation on C&I focuses more on generating data than on its analysis. This is where the ACM teams sought to complement the framework with another perspective.

Analysis Reflection and Adaptiveness: Participatory Action Research

In many experiences with participatory monitoring and evaluation, the starting point was often an externally managed project or program in which participation meant little more than involving community members in identifying indicators or collecting data. Little attention was paid to the reflective and analytical processes. This was partly based on a commonly held assumption that the use of indicators, as long as they were developed based on local perceptions and information needs, would lead to learning. But it was also driven by lack of awareness about the importance of investing in effective reflection.

For information to make sense and be useful, it must be analyzed. Participatory action research offers a roadmap for critical reflection because it combines change and the pursuit of insights through "a cyclic process that alternates between action and critical reflection" (Dick 1999). PAR can be defined as "collective, self-reflective enquiry undertaken by participants in social situations in order improve the rationality and justice of their own social . . . practices" (Kemmis and McTaggart 1988, 5). It consists of four steps—reflection, planning, action, observation—that follow each other cyclically and are interdependent (Figure 1-3). Because the process is so varied in practice, it is perhaps better to talk about a spectrum of research approaches, from "traditional" research at one end to "participatory research" at the other (McDougall and Braun 2003); PAR is clearly at the participatory end of the spectrum.

Within the context of ACM, participatory action research can operate simultaneously at two levels. It can be used to help implement and adapt the resource management regime, but it can also help develop and implement monitoring mechanisms[10] that fuel the adaptations.

In the first case, a PAR-based approach is particularly relevant when the situation involves many unknowns and diverse perspectives and when the participation of those affected by the change is critical if subsequent action is to be effective and sustainable. This is typical of the forest areas and resource problems in ACM research cases. Forest management actions are viewed as experiments about which learning is required through explicit deliberations that in turn inform new actions.

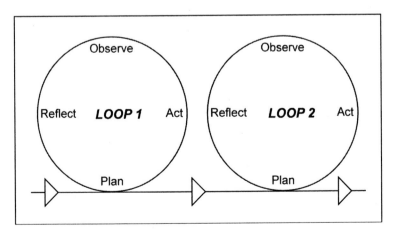

Figure 1-3. *Participatory Action Research Cycle*

At each step, understanding becomes clearer and in turn allows for improvements in methods, data, and sense making.

In the second case, PAR can be used to shape collaborative monitoring structures. It asks stakeholders to be clear about a shared concern or question that needs tracking, be it a long-term vision of their ideal forest (Chapter 3) or a focused problem, such as harmful beekeeping practices (Chapter 10). They jointly develop a systematic plan to deliberately assess their actions as an instrument for achieving ACM. Then the information they have sought is scrutinized and analyzed to determine next steps. Participants in ACM create and adapt their own learning processes following an action research cycle.

Because of PAR's focus on a conscious and reflexive cycle—plan, act, reflect—the monitoring process itself can continually be scrutinized and improvements made in methods and definitions of useful information and sense making. Several chapters in this volume discuss the adaptiveness of collaborative monitoring (Chapters 10, 11, and 12) as the users gain insight into how to make the best use of monitoring. Given the many unknowns related to monitoring, a PAR perspective thus allows collaborative monitoring to evolve through continual reflection.

In practice, PAR calls for an attitude of curiosity and questioning (Box 1-3). This often requires fostering by facilitators (Chapter 11). Rarely, however, is critical thinking consciously encouraged such that participants dare to question and critique (Klouda 2004). Although the facilitator's approach and attitude are essential to making this possible, specific methods and tools can also be useful for collaborative analysis. Several chapters describe such methods and their use (Chapters 3, 4, and 10).

Supporting the Process: Facilitation

Forests are known "arenas of conflict" between the many social groups with diverse and often competing claims on forest resources (cf. McCarthy 2004; Li 2002; Biesbrouck 2002; Engel et al. 2001; Ghimire and Pimbert 1997). To manage conflicts and negotiate next steps, forums are needed and sometimes may exist. A useful notion when thinking about such forums is that of "space," or "opportunities, moments and channels where citizens can act to potentially affect policies, discourses, decisions and relationships which affect their lives and interests"[11] (Gaventa 2006, 5). Cornwall (2002) differentiates between three kinds of spaces—closed, invited and created spaces[12] (also see Chapter 13)—stressing that in each, relationships of power determine the quality of people's participation.

Much forest management continues to take place in so-called closed spaces, with decisions made by elites or elected representatives but without open access or possibility of wider societal debate. ACM marks a move away from such closed-space forest management. Those engaged in ACM seek to create opportunities for participation by marginalized forest users. Ensuring forest equity requires creating space for the less vocal to have their say or using invited spaces for the benefit of the marginalized. However, decision-making power is rarely relinquished voluntarily by vested interests; it must be (re)claimed and negotiated continually.

Box 1-3. Understanding and Ensuring Truly Critical Reflection

The term *critical reflection* is often used without a full understanding of its meaning. Learning comes from realizing that there is a difference between what one expected to happen and what actually happened, and then identifying which assumptions might need revising and how. Reflection relies strongly on the ability (and the opportunity) to challenge the assumptions that informed the actions. Hence reflection can be considered the way in which people explore and reassess their assumptions.

To reflect, a person needs a certain level of curiosity that enables her or him to "flick between inquiring and interpreting before developing useful insights ... three things help the reflection process: challenge—the puzzle needs to be tough but not too tough; pragmatic curiosity—a real puzzle with a real purpose for resolving it; and the issue needs to be slightly fuzzy." (Bob Williams, personal communication, summarizing views of Siebert and Daudelin 1999).

Normally, projects pay relatively little attention to critical reflection. Most monitoring efforts and resources are spent on collecting data, statistical analysis, and basic reporting of activities—for several reasons. Data collection is more mechanical and requires less analytical capacity. Monitoring is often seen as a bureaucratic obligation stipulated by the contract and by government regulations, and not necessarily of value to the initiative. Few people seem to consider frank analysis of underperforming initiatives as worthwhile. Reversing such views requires building capacity and creating effective incentives.

Many factors impede learning from critical reflection in organizations. For example, John Edwards found that in Australian organizations, reflection and learning are regarded as a luxury that "gets in the way of work" and must therefore be undertaken in one's own time. He also found that people are not permitted to admit ignorance and are under considerable pressure to provide quick solutions.

Source: Based on Bob Williams, personal communication.

Facilitators[13] can play a significant role in either helping create spaces or making invited spaces useful for marginalized groups. The facilitation of agricultural change processes is accepted (cf. Hagmann 1999), but it is also needed for the monitoring of ACM. Many chapters in this volume describe how the quality of collaboration in monitoring benefited significantly from the presence of facilitators. In Nepal, the facilitators took existing processes and mechanisms as their starting point and introduced new considerations, such as critical reflective capacities and gender equity. Santos and her colleagues in Brazil took on a stronger facilitation role because collaborative monitoring was an innovation.

Facilitators can negotiate and maintain opportunities for collaboration in monitoring, performing a range of roles. Mutimukuru and her colleagues (Chapter 11) describe how the facilitation team worked with both sides—local forest users and government forest guards—to create a new and shared understanding of what collaborative monitoring would mean. Kusumanto's work also involved working with multiple interest groups to ensure an open local election process and, perhaps more importantly, to support local shifts in norms of what constituted a fair election. By contrast, in Porto Dias, Brazil, the facilitators chose to work mainly with rubber tappers, whose perceptions of forest management were too divergent to be integrated with those of recent settlers at the first stage.

Another role for facilitators is that of go-between. In the Philippines, Hartanto found that the facilitators acted as useful mediators, able to question existing systems, between the powerful municipality and the less organized local people's organization. Oyono and his colleagues in Cameroon (Chapter 9) served mainly as monitors of the monitoring process. In almost all the experiences, facilitators provided methodological ideas and options, some building on local systems (such as Cronkleton and his colleagues, Chapter 5) and others bringing in the more comprehensive C&I framework (such as Pokorny and colleagues, Chapter 2). In all these interactions and roles, facilitators become part of existing dynamic relationships of power and bring to it their own power. It is critical for them to be conscious of their position and how this can open opportunities and influence the monitoring process and the results.

Diversity of Context, Diversity of Practice

The three activities—using indicators to structure information collection, PAR to guide critical reflection and adaptation, and facilitation to mediate the process—have shaped the experiences with collaborative monitoring discussed in this volume. But despite this common starting point for each of the ACM research teams, contextual differences led to unique learning processes. Maarleveld (2003) suggests that four questions influence how learning processes are constructed:

- Why invest in learning?
- Who should be involved?
- What should we learn about?
- How do we create learning opportunities?

Principal among these questions is the first one—that of the core purpose—since it determines subsequent choices. The learning intentions will indicate which stakeholders to involve and their roles: who initiates the process of conscious and ongoing reflection, and who makes other essential contributions. The selection of stakeholders is often shaped early on, with only limited scope for adjustment at later stages. This may require patience with limitations in the initial phase followed by ongoing attempts to push the boundaries of who may be included and the conditions of their engagement. Also critical are the scope and focus or level of monitoring efforts, which are strongly shaped by prior history of the management interactions. The fourth question requires careful thought about the three building blocks discussed above. The remainder of this section focuses on the first three questions.

Purposes of Monitoring

A common tendency in collaborative monitoring efforts of any kind, including in the forestry sector, is to try to fulfill diverse monitoring purposes with a single information collection and processing system and through a single analytical process, rather than mutually complementary but distinct efforts.[14] People want to provide information to funding agencies, contribute to local empowerment, make technical assessments of forest management, and ensure equity.[15] A crisis or phenomenon, such as the transition to local control of forests, might need tracking (Chapter 9). Innovation in resource use might be needed and its effectiveness tracked (Chapters 3, 5, and 10). These are all valid monitoring purposes. However, information collection and decision making become confusing if they are all wrapped up into a single system to which all the different stakeholders must relate.[16]

Each monitoring purpose has a different audience with specific information needs. Therefore, each will have a distinct cycle or rhythm appropriate to stakeholders with specific roles and a unique information focus. For example, assessing the effectiveness of the new honey production regime in Chimaliro will require long-term continual tracking and public debate (Chapter 10), whereas Kusumanto's example of monitoring an election process in Indonesia is more time bound (Chapter 12).

The monitoring systems in this volume were created to meet different purposes. In two cases, collaborative monitoring was intended for and contributed to equitable access to the resources (Chapters 3 and 4). This single purpose led to highly focused systems. McDougall et al.'s example cites increased equity as one aspect in a wider attempt to enable marginalized voices to be heard in community forestry management in Nepal (Chapter 8). Groups may choose to develop a single common framework for observing the effectiveness of their plans and the unexpected outcomes (Chapter 7). In other initiatives, a subgroup may agree on what to monitor and how to use the information for its own planning but also in negotiations with other stakeholders (Chapters 3 and 5).

Thus, in collaborative monitoring, various purposes (co)exist, ranging from data gathering for upward accountability to funding agencies to joint analysis for learning what improvements might be needed. All are needed but each may require specific mechanisms and participants. Answering the question "Who needs to be learning what and why?" may help in identifying the core purpose(s).

Participants and Their Roles in Monitoring

Adaptive collaborative management involves a rich mix of individuals and groups who affect forest resources and share multiple and interconnected stakes in these resources. The role that individuals and groups play in monitoring their ACM initiative can be that of initiator, driver, implementer, analyst, or audience—or various combinations. Choices made in relation to this question create some of the diversity in collaborative monitoring.

One role that is critical for the effectiveness of monitoring efforts is that of "driver," the group that sustains the momentum of the initiative. Understanding who is driving the development of the monitoring system can give clues about what is needed for sustainability. This may not be the ideal group to sit in the driver's seat. If it is an external group, then local ownership may be jeopardized in the absence of a handing-over strategy. If it is a local group, then a careful look at on-site methodological capacity is needed. With their insights about power relationships among the actors involved in the monitoring, facilitators can address inequities by putting the question "Who *should* drive the monitoring?" on the agenda.

Many chapters in this volume demonstrate an awareness of the need for local ownership if monitoring is to strengthen ACM. Cronkleton and his colleagues (Chapter 5) describe how they, as external facilitators, focused on an immediate concern—monitoring of benefit sharing—to embed the idea of monitoring as "good practice" and one that could be developed further locally. Nyirenda and Kozanayi (Chapter 6) describe similar concerns for local ownership—a locally initiated and locally driven monitoring system for broom grass groups. Then they reflect on how it may serve as a useful entry point for externally facilitated expansion of the monitoring into new topics.

A new role for many at the community level is that of analyst and researcher. Pokorny et al. describe working with local stakeholders to prepare them for these roles and accompanied them during the research and analysis in strengthening these new skills. In both Nepalese examples (Chapters 4 and 8), the *tole* (hamlet) took on a new monitoring responsibility for generating debate, making decisions, and raising concerns at the village-wide meetings. Those working at the *tole* level had to grow into these new roles. The importance of capacity building is discussed in more detail in the concluding chapter.

Scope and Focus of Monitoring

A third aspect that creates diversity of monitoring practice is the "what" question. Specifying the monitoring content occurs after the purpose(s) and stakeholders are clear. Two dimensions need consideration: what is the scope and what will be the focus of the monitoring effort?

The scope of monitoring will be affected by the scope of the ACM initiative itself. Is the forest management regional, community based, or organization focused? How many people are involved, what are their relationships, and how extensive is the ACM experience? If the scope is large but capacities are limited, then perhaps the strategy will be to start small and focus the monitoring on one level or one theme. If the scope is limited, then diverse issues or levels of monitoring might be feasible.

The scope of collaborative monitoring is also subject to change over time. Initial efforts may be limited, but as capacities grow and information needs shift, the systems can expand. Kusumanto, Cronkleton and his colleagues, and Kamoto all describe how this happened in their experiences (Chapters 5, 10, and 12).

Focus derives directly from the topics of prime interest. Is the ACM effort focused on creating shared meaning and coordinated action, improving forestry performance, enhancing local livelihoods, or challenging assumptions of collaborative forest resource management?

Sometimes focusing on a microissue can be effective; examples include income in Bolivia (Chapter 5) and broom grass harvesting in Zimbabwe (Chapter 6). McDougall and her colleagues describe another focus, equity, that emerged in their work in Nepal (Chapter 8). Meanwhile, in Indonesia (Chapter 12), the focus was also consciously narrow to allow time for building local capacities and interest in collaborative monitoring.

Monitoring can also be limited or more comprehensive, touching on various forest resources and aspects of their use. The information being sought can relate to the quality of social interactions, prevalence of resources, harvesting of resources, and so forth. In the Philippines (Chapter 7), monitoring diverse issues was made more feasible by sharing responsibilities among stakeholders. In Brazil (Chapter 3), a wide range of resource issues was monitored but narrowed down to a few stakeholders and municipalities. Information needs can also be quite specific and concern only beekeeping (Chapter 10) or broom grass (Chapter 6).

Both scope and focus will be strongly shaped by the history of the ACM initiative and by its evolution. A new initiative may require a simple entry point (Chapter 5) that can expand with changing information needs (Chapter 6). However, a new initiative may also justify a more comprehensive approach (Chapter 3). An older initiative may benefit more from focused efforts around a weak aspect, such as equity (Chapter 8), or it may have the capacity to manage a more elaborate effort. Amid the diversity and options, what is critical is that scope and focus are the logical result of the choice of purpose and stakeholders.

Structure of the Book

This book is structured in four thematic clusters. The first concerns the criteria and indicators logic that initially drove CIFOR's work on adaptive collaborative management. This section discusses the complexity of implementing CIFOR's framework for monitoring forest systems. Pokorny and his colleagues (Chapter 2) explain what they attempted in the eastern Amazon of Brazil to make the framework more inclusive and meaningful to local communities. They concluded that in their approach, the framework provided more external benefits to scientists than local benefits. In a different part of the Amazon, Santos and her colleagues (Chapter 3) applied the C&I framework with rubber tappers to sustained production of timber using low-impact technologies for selective logging. They too struggled with making it participatory and empowering and sought to make the system simpler and more agile.

The second theme is building on local dialogue, tracking, and learning systems. The first case, from Nepal, reveals how Paudel and Ohja slowly readjusted their initial assumptions to align with local realities (Chapter 4). They saw how their initial focus on indicators to assess planned activities detracted from other, more important tasks to enable local participation in forest management. Cronkleton and his colleagues in Bolivia (Chapter 5) discuss two communities where monitoring was used to track benefit sharing from logging concessions and explore why one case was more successful than the other. The third contribution in this section, from Zimbabwe (Chapter 6), also compares several local resource management committees and notes the effectiveness of one case that was simple and locally relevant.

The third cluster presents reflections on how to deal with sociopolitical differences when creating monitoring systems to strengthen collaborative action. Hartanto talks of Palawan, the Philippines (Chapter 7), where differences in information needs and, above all, power between government officials and the local resource users caused withdrawal of the government. They stress the need to build local capacity to renegotiate monitoring agreements and arrangements. McDougall and her colleagues in Nepal (Chapter 8) tell the inspiring story of how active involvement in developing a monitoring process enabled one forest user to understand her rights and then hold the group accountable to its own equity-related goals.

The final theme focuses on ensuring adaptiveness of monitoring to accommodate changed needs, new actors, and increased capacities. Kamoto's work in Malawi (Chapter 10) illustrates the evolution of the system as insights grew about what benefits the monitoring can bring—but also what hitherto accepted behavior was now being questioned. Mutimukuru and her colleagues in Zimbabwe (Chapter 11) share a range of powerful lessons, discussing among other issues the importance of starting small before expanding to incorporate more stakeholders with divergent interests. Kusumanto rounds off this theme with a detailed example of how one local community evolved through facilitated reflections toward a situation in which the elite-dominated decision making was broken, and transparent and accountable elections could be held.

The closing chapter synthesizes the insights gained by the contributors about the factors that shape the design process and determine implementation of collaborative monitoring in natural resource management. Despite its practical focus, this book is not a manual. It does not describe a single guaranteed model of how to develop, support, and implement collaborative monitoring in forest areas. It does, however, present a variety of experiences, some that have fed management improvements and others that have yet to bear fruit. They all offer methodological insights about what seems to work, what hinders progress, and what to look out for. In so doing, they can guide others working in community forestry. More importantly, perhaps, the authors discuss issues critical for anyone embarking on a collaborative monitoring route in natural resource management. Together, the articles draw principles from practice that, with creativity and commitment, allow for adaptation in diverse settings.

Notes

1. Various dimensions were analyzed by Colfer (2005a): the extent to which authority over natural resources had devolved to local communities; forest type and its perceived relative quality; population pressure; stakeholder diversity and the related multiplicity and divergence of management goals; the level of conflict; and the existence of social capital.

2. ACM-related publications to date have looked primarily at planning and implementation with multiple interests (e.g., Edmunds and Wollenberg 2001; Wollenberg et al. 2001a); governance (e.g., Sarin et al. 2003; Wollenberg et al. 2003; Diaw 2002); equity considerations (e.g., Upreti 2001; Colfer 2005b); modeling options (e.g., Haggith and Prabhu 2003); impact levels (e.g., Colfer and Byron 2001; Colfer 2005a); scenarios (e.g., Nemarundwe et al. 2003); and participatory research processes (e.g., McDougall et al. 2003; Sithole 2002; Colfer 2005a). ACM

work on monitoring has focused almost exclusively on indicators (e.g., Mendoza et al. 1999; Ritchie et al. 2000; CIFOR 1999a, 1999b).

3. This section draws heavily on Colfer (2005a, Chapters 1–3, pages 1–50) and various ACM documents, notably the International Steering Committee reports and official project documents (cf. Prabhu et al. 2002).

4. This concern is central to all CIFOR's ACM research and related publications, including Hartanto et al. (2002), Colfer (2005a, 2005b), Kusumanto et al. (2005), Diaw et al. (forthcoming), Fisher et al. (forthcoming), and Matose and Prabhu (forthcoming).

5. CIFOR's interest in collaborative monitoring is part of a growing interest in participatory forms of monitoring and evaluation in all sectors and regions since the mid-1990s (cf., Estrella et al. 2000; Baron 1998; MacGillivray et al. 1998; Abbot and Guijt 1998; Herweg and Steiner 2002; Ottke et al. 2000).

6. The fourth activity—applying the information to derive lessons—has not been treated separately by the authors but is embedded within the participatory action research process.

7. This was one of three sets of C&I, the other two being for commercial timber (CIFOR 1999c; Prabhu et al. 1999) and plantation management (Muhtaman et al. 2000; Sankar et al. 2000).

8. These are community-managed forests whose flow of goods and services can be maintained without reducing their quality or value for future generations (Ritchie et al. 2000, 1).

9. Prabhu et al. (1996) compare this hierarchy with one that is perhaps better known, namely wisdom, knowledge, information, and data.

10. Information frameworks (such as the C&I hierarchy discussed above) are a part of these mechanisms.

11. The notion of space was not part of CIFOR's ACM concept.

12. Closed spaces are defined as official or unofficial spaces to which only certain people or interest groups are invited, and in which others are excluded; invited spaces are formal or informal spaces in which powerful officials invite people or organizations to be consulted or to make their views known; claimed spaces are formal or informal spaces created by those who seek to have greater power and influence (Cornwall 2002).

13. In participatory action research, ACM team members are researchers and facilitators simultaneously.

14. For example, Ottke et al. (2000) discuss a range of purposes, from self-assessment of performance to monitoring forest cover for advocacy. Yet they have a single set of steps that they believe can deal with various purposes at the same time.

15. The concluding chapter in this book discusses the range of monitoring purposes in more detail.

16. See Table 13.2 in Chapter 13 for more details about diverse monitoring purposes.

PART II

Using Criteria and Indicators

CHAPTER 2

Testing the Limits of Criteria and Indicators in the Brazilian Amazon

Benno Pokorny, Guilhermina Cayres, and Westphalen Nunes

C RITERIA AND INDICATORS (C&I) TO ASSESS sustainability as developed by CIFOR are tools designed to deliver information required to conceptualize, evaluate, and implement sustainability (Prabhu et al. 1998a) (see Table 1-3, Chapter 1). They denote a hierarchy of linked items (principles, criteria, indicators, and verifiers), such that the information accumulated at lower and more concrete hierarchical levels is used to assess the related items of the upper, more abstract levels (CIFOR 1999a). The C&I framework represent a type of communication network that allows the different actors involved in forest management to discuss the requirements for sustainability and to inform one another about progress toward the goal. Since the 1992 Rio "Earth Summit," whose declaration identified the concept of sustainable forest management as a promising option for maintaining tropical forests (UNCED 1992), various sets of criteria and indicators have been developed to help evaluate progress toward more sustainable forms of development at the national scale and in commercial management units. Significant efforts have also been undertaken to address the needs and demands of small producers within the criteria and indicators (IUCN 1997; Colfer et al. 1999a, 1999b; Ritchie et al. 2000; Pokorny et al. 2003a).

Due to their potential utility in disaggregating complex issues into smaller communicable elements, criteria and indicators are considered helpful in facilitating communication between stakeholder groups involved in natural resource use, and therefore in providing a methodological basis for collaborative learning processes about resource use. Although C&I were originally designed for monitoring sustainability at the national level and in forest management units (e.g., ITTO 1992; CIFOR 1999a; TCA 2001), increasing numbers of initiatives have begun to recommend criteria and indicators to rural communities for their own, local use in participatory resource management processes. Some of these initiatives have even highlighted the criteria and indicators themselves as the most significant benefit for communities. Promoted by international organizations such as CIFOR, Forest Stewardship Council, and the Worldwide Fund for Nature (WWF), computer software has been developed to support the use of criteria and indicators at the smallholder level, with the expectation that this would help make their development and application a locally embedded process.[1]

However, our experiences throw up considerable doubts about the expectation that criteria and indicators might provide great benefits to local communities. Field workers using C&I-based methodologies, notably for local monitoring systems, have reported various difficulties. In particular, they say that criteria and indicators as promoted by their organizations do not fit the specific conditions, interests, and capacities of rural people and should not be applied in local resource management processes.

This chapter presents two ambivalent experiences from the Brazilian Amazon with the use of C&I as suggested by CIFOR (CIFOR 1999a; Ritchie et al. 2000). Together, our experiences lead us to suggest a more realistic expectation about the opportunities and potential benefits of criteria and indicators for collaborative learning processes with rural communities.

Experience 1. C&I for a Collaborative Background Study

The first experience comes from 1999, when CIFOR's regional adaptive collaborative management (ACM) group started working with three communities in the eastern Amazon as part of a research project about collaborative and adaptive forest management (Pokorny et al. 2003b)[2]. The ACM group included five researchers, specializing in forestry, sociology, agriculture, linguistics, and geography. The three communities are about 200 to 400 km from Belém, the capital of the State of Pará (Figure 2-1). The *ribeirinhos,* as the inhabitants of Amazonian riversides are known, operate in traditional communities characterized by a subsistence economy that is dominated by products of the Açaí palm (*Euterpe oleracea* Mart.).

As part of the methodological approach, the research project called for a background study to obtain some basic understanding about the communities and subsequently enable comparative analysis (cf. Colfer 2005a). Normally, external researchers would carry out such a study, along with a stakeholder analysis and historical research, and would use criteria and indicators to help structure and analyze the information. However, a traditional background study uses communities as a source of information for ongoing research and thus clashes with the participatory intent of adaptive collaborative management (Chapter 1).

Believing that the information needed for the background study should also be important for the communities, we decided to invite active community involvement in the C&I-based background study. We expected that this could significantly support the process of sensitizing and empowering the communities in dealing with the concept of sustainability and enrich their perception of their ecological and socioeconomic environment. Thinking about how to make the background study their study led us to the idea of a collaborative diagnostic study, which would include the collaborative assessment of the C&I set defined by experts as a basis for

Figure 2-1. *Eastern Amazon, State of Pará, Brazil*

later adoption. We also expected that the communities would then critically evaluate the C&I that had been defined externally and use them as a basis for defining their own criteria and indicators for local monitoring.

Modifying the Tools

In a first step, the project staff defined the C&I set to be assessed by the community and immediately encountered difficulties. The system of interrelated items at different hierarchical levels that lies at the heart of the C&I approach proved challenging. The researchers had difficulty defining items with a similar degree of specificity for the different hierarchical levels and aggregating them into items of higher hierarchical levels. Furthermore, they were unable to decide about the feasibility and relevance of the defined indicators because of a lack of field experience with the C&I approach.

To help develop the C&I set, the team tried to use CIMAT, a software designed by CIFOR to help users adapt criteria and indicators to meet local conditions and expectations. However, the team found that the time-consuming interaction with the computer broke the spontaneity needed to respond adequately to new ideas for structuring and organizing the C&I. It became obvious that especially in the beginning, a group needs to play flexibly with the criteria and indicators before making a final decision about how to organize them. Most critical was the fact that only one person had direct access to the computer, and the others felt excluded and marginalized. Therefore we decided to abandon CIMAT and defined a preliminary set of criteria and indicators in a more conventional manner by using cards: one idea was written on each card, and the cards were then sorted and clustered on tables and walls until agreement was reached.

Our next step was to shift this process to the communities in the form of preparatory workshops. There, when trying to convey the concept of C&I to participants, we encountered immense difficulties. The terms *criteria, indicator,* and *sustainability* created confusion and skepticism. Although we worked with examples—and these were understood—all participants had serious problems condensing information at the more abstract levels of criteria and principles. Nobody really understood the reason for undertaking such a complicated endeavor. For the participants, the detailed information at the verifier level was sufficient for feeding the discussion and communicating ideas. In addition, the large number of indicators they identified were often worded in ways that caused misunderstandings. The problem was aggravated by the high level of illiteracy and discomfort in working with written material. Thus, we decided not to discuss criteria and indicators directly but to focus on the substantive issues related to forest resources. In so doing, we hoped to receive information needed to understand local views better, and we would then translate their perceptions into the language of C&I.

To enable a discussion about this content, we classified the C&I into thematic clusters. Based on the recommendations given in CIFOR's C&I Toolbox (Colfer et al. 1999a, 1999b), we selected a set of participatory methods for each of these thematic clusters to discuss criteria and indicators with community members. These methods were used during a sequence of workshops held in the evenings to accommodate participants' schedules.

In the most intensive study, conducted with the Canta-Galo community in the district of Gurupá, we held nine workshops during three weeks, each workshop treating the content of one specific criterion. Despite this intensity, all workshops were attended by nearly the whole community, which to us seemed to indicate their interest and motivation. In the other two communities, the level of attendance varied from 5 to 30 people. Obviously, in these communities, most people were interested enough to be involved in the discussion process but not sufficiently to invest their free time in the evening. The level of interest had no clear relation to the social status or gender of the community members or the subjects discussed.

In each community, we tried to use the CIMAT software again to document the outcomes of the group discussion during the workshops. Again we found the software inflexible, unwieldy, and inadequate for visually conveying the hierarchy of concepts. We decided to use a simple ACCESS database instead but still faced several logistical problems. The simple lack of tables and chairs hindered the work with computer equipment. Another problem was the unstable electricity supply: the diesel generators in the villages operated for only the early part of the evening. To avoid damaging the hardware, it was necessary to use stabilizers, which were heavy and extremely sensitive to humidity and shock. We considered using solar cells but these proved too expensive.

The laptops did create immense interest in the community. During the computer work in the evening hours, many people came together to assist with the database work. However, it was impossible to make full use of this high level of interest. First, the laptop screen was prohibitively small. Second, the local participants did not know how to handle the software, which made intensive training necessary. Ultimately, the main reason for abandoning the use of computers was the technology bias and the exclusion it implied. Community members were not involved in entering and processing the data and so did not use the information—yet they were the very people who could benefit from any insights gained.

Questioning the Local Value of C&I

Although the workshops revealed interesting insights into local views and understanding of the social, economic, and ecological environment and generated important information for the background study—and thus helped us organize and articulate the criteria and indicators—we did not see the communities make much progress in understanding and using C&I. They were less interested in developing a systematic set of criteria and indicators than in discussing specific issues directly relevant to their daily lives. A core dilemma of the C&I work emerged clearly—between the comprehensive and systematic evaluation based on CIFOR's C&I logic and the selective and concrete information needs of the communities.

In our attempts to assess the entire C&I set, we reverted to an externally dominated process of obtaining information, despite our participatory intentions. The sophisticated technical characteristics of the C&I framework and local participants' intellectual distance in formulating C&I necessitated strong leadership by outsiders. External dominance, however, neither supports empowerment goals nor respects local values and views sufficiently. Consequently, our initial idea to support communities in assessing their own C&I was not a success. For them, evaluating the complex issue of sustainability by using a full set of C&I was too abstract and time consuming. Based on this experience, we concluded that the potential of a C&I framework as a communication tool in collaborative learning approaches at the community level provided benefits for outsiders but not for the communities.

We did notice that the participants' communication skills improved during the study. They became more self-confident in expressing and discussing personal opinions and controversial aspects even with us, as outsiders. Their ability to work effectively in groups improved. The general evaluation of the study by the community was very positive.

Nevertheless, this should be interpreted more as a result of working with participatory methodologies rather than as a specific outcome of dealing with criteria and indicators. From our observations, we suspect that other methodologies may be more appropriate for discussing issues of specific relevance or working on a few very relevant indicators (Chapter 6 and 10) than trying to assess sustainability by using a complete set of criteria and indicators.

Experience 2. Participatory Assessment of C&I

Despite the limitations of our C&I work at the community level, the effort had a positive result for us as a project team. Developing a C&I set had provided us with an integrated understanding of our planned activities. It enabled us to perceive and discuss our different viewpoints and biases, to start a critical evaluation of our observations and information, and to identify and accept (differing) community opinions and values. This gave us several ideas about how trained experts could work with this instrument. For example, we thought of using C&I sets as an agenda for an integrated assessment of living conditions relevant for local human well-being and maintenance of environmental services.

The authors of this chapter, who had been involved in the experience in the eastern Amazon, tested this idea in a project[3] about opportunities for local farmers in the northeastern part of Pará, Brazil, which sought to enhance human well-being and environmental stability by managing secondary forests. We proposed that participating families assess a list of indicators as a way of developing an integrated understanding of their living conditions and how these conditions influence the management of their secondary forests, and vice versa. We expected to generate a better understanding about how factors within and outside their direct control influence their resources, as a basis for discussing opportunities to improve their living conditions and engage in political processes (Nunes et al. 2004).

Based on a C&I assessment of a forest management unit by four stakeholder groups (Pokorny et al. 2004), we designed a five-step process: (1) mobilization, (2) preparation, (3) field assessment, (4) analysis, and (5) presentation of results. Using the C&I set developed during the collaborative background study discussed above, the project team compiled a list of 159 indicators that could be assessed easily and were relevant for the sustainability of secondary forest management initiatives. We then classified these 159 indicators into seven groups: basic needs, legislation, public policies, organization, culture, forest management, and economy and ecology.

Mobilization

In step 1, we presented our suggested methodological approach and a possible time schedule during workshops in selected communities organized by the project in three municipalities. All municipalities are characterized by relatively longstanding settlements, and thus the rural population is dominated by small farmers managing 20- to 50-hectare plots. Our objective was to identify about three to five people in the selected communities of each municipality who were willing to invest about two weeks of volunteer time during the three-month study period. We said that no remuneration could be expected, except food during the workshops, and that the main benefit for participants would be gaining insights into their situation and learning how to find information about critical issues in their lives. We were surprised that despite the significant time commitment, more than 35 small-scale farmers from 10 communities were interested in participating. Most were motivated by the possibility of learning how to access information, as well as by the hope of learning how to improve their precarious living conditions.

Preparation

We invited all interested farmers to a three-day preparatory workshop in Belém, the capital of Pará. Our main aim was to explain the concept of working with indicators and to prepare the field assessment. The farmers, organized in working groups, defined what information was needed from which sources to assess the 159 indicators in our seven indicator clusters. Through this exercise, we hoped that the participants would obtain an overview of the C&I set and better understand the overall task. We worked in small groups of four or five participants to ensure active participation of everyone.

However, each group had difficulties with the exercise, tending to assess whether the indicator was good rather than determining what information each indicator needed. We had to explain the indicators and group tasks repeatedly. Literacy problems and participants' limited experience with group work made the process time consuming, with each group needing more than half an hour to discuss one indicator. Nevertheless, with strong support by facilitators and by extending the workshop for one day, the groups managed to define the information needed for the assessment.

During this workshop, we tried to use a simple computer tool to facilitate the systematic documentation of group work and preparation of the field reports. However, none of the participants were experienced with computers, and those participants not directly involved in the data processing felt excluded. Thus, all groups decided to work only with cards.

Field Assessment

For each of the three municipalities, one data collection group was formed, and a facilitator was assigned to each group. In each municipality, the same set of indicators was to be assessed and therefore the same information was to be collected. The first activity for each group was to define a detailed work plan, including tasks per group member.

Three days were set aside for the data collection. The information was collected mainly through interviews with different stakeholder groups, such as other communities, bank directors, and representatives of the municipal government. Before each interview, the groups identified the indicators relevant for the interview and checked what information was needed. This helped them prepare for the interview and document the information related to each indicator.

After each interview, the group discussed how they had performed as interviewers. Through these group evaluations, they developed a greater sense of self-awareness and rapidly managed to improve their efficiency. Normally, the information gathered during the day was discussed in the evening. After reviewing the information, the group decided whether they had enough for each indicator.

The participants identified several benefits from this fieldwork. They gained insights into their institutional environment and learned to overcome communication barriers with other stakeholder groups with whom they had not previously been in contact. They also increased their understanding of the causes of the problems in their lives and learned about the difficulties and opportunities for improving the situations and about the complexity and interrelation of political, social, economic, and environmental issues.

Analysis of Findings

After the field assessment in the municipalities, the three groups came together in one workshop to compare their results and identify the similarities and differences among the three municipalities. Discussing why an indicator in one district was assessed differently in another proved to be an excellent starting point to analyze underlying causes. The overall consensus from the C&I assessment was that the situation with respect to all seven indicator clusters in the three municipalities was precarious.

The groups were also able to identify various opportunities to improve their situation. Some of the more promising opportunities included improved organization of their own communities, increased participation in unions, more intense dialogue with politicians, and more careful selection of their political representatives.

During the workshop, participants preferred to reflect on what they perceived as major problems instead of systematically discussing all 159 indicators. They perceived their findings as being the result of intense group discussions rather than systematic data collection. Thus participants concentrated their discussion on the 54 most relevant indicators.

Our attempt to condense the assessment to the level of the seven indicator clusters failed because the group focused exclusively on detailed examples, personal experiences, and concrete situations. Again, the hierarchical system of C&I and the related possibility of systematically aggregating and disaggregating information was not accepted.

Presentation

In the last step, the groups presented their findings, first separately, in their communities, and then together at a more public event to which political representatives of each district had been invited. The events were organized by the project, but the groups made the presentations themselves with the support of a facilitator. In their presentations, the groups focused on the principal problems and opportunities. None mentioned the term *indicator* or the underlying approach. They later said it would have been too difficult and time consuming to explain this concept to "outsiders."

After the five-step assessment process was finished, we encouraged some participants to document their views by revising the indicators to provide a basis for similar studies and to enhance the consideration of local understanding in political discussions. Documenting the C&I work would help them communicate with other important players. However, even after their intense experience with criteria and indicators, the participants were unable to make specific and constructive suggestions for the indicators. Comments remained very general, such as about the excessive number and exhausting work. They generally confirmed the utility of all indicators defined by the project team and suggested that future indicators be more specific in content and simpler in wording. In summary, referring to the terms in CIFOR's C&I framework, a methodological discussion was possible only at the level of verifiers that related directly to their lives and perceptions—and not indicators.

Lessons

The criteria and indicators framework for communicating about the sustainability of natural resource management can potentially provide important inputs for collaborative learning processes. However, our experiences show the need to evaluate how to use them at the level of small local producers and rural communities. It is clear that C&I are a so-called expert system: proper use of criteria and indicators requires specific knowledge and intensive training of technicians and scientists, and therefore of any other group, such as rural people, who might be involved. If local people are to be included in the decision-making processes that criteria and indicators are supposed to support, then training is essential and will require considerable input and dedicated interest of all those involved. However, it does not appear justified to train entire communities in understanding the complex concepts of C&I for the assessment of sustainability.

For communication and evaluation processes inside communities, we concluded that criteria and indicators are not adequate. The logic of the C&I approach, which is based on a complex system of hierarchically linked items, does not correspond to the capacities of local people, whose way of thinking is characterized more by concrete individual perceptions. In the language of C&I, local people tend to focus on the level of verifiers. Furthermore, they simply do not have the time for a more systematic application of complex indicator sets. Consequently, the use of criteria and indicators frameworks, such as CIFOR's, in rural development projects will probably remain in the hands of outsiders and thus not contribute to local empowerment processes. However, as mentioned above, working with a few very relevant indicators as a basis for monitoring and planning is a different approach that shows a high utility for local communities.

The same tendency for only marginal local benefits was detected in the application of computer tools, which are being simulated to support work with criteria and indicators. Even if local people are properly trained to use this medium, they still depend on outsiders for equipment and supervision. Positive experiences require dedicated and continuous external support. Most critical is our observation that using computer tools can create insiders and outsiders within a social group. Using materials, mechanisms, and knowledge already available in the communities, on the other hand, can diminish dependence and outsider dominance. The simpler, the better. That local people can and should become qualified in modern technologies is extremely important, but our experiences suggest that this should be separated from the use of participatory approaches.

Notes

1. Examples of such software are CIMAT, developed by CIFOR, and Pathfinder, developed by Forest Stewardship Council/WWF/IKEA.

2. The described experience was also part of a project to evaluate the compatibility of C&I sets used in the Brazilian Amazon, financed by the German Development Organization (GTZ).

3. This project was financed by ProManejo, one of the subprograms of the International Pilot Program for the Protection of Brazilian Forests (PPG-7). The project took place as a scientific cooperation agreement between EMBRAPA (the state research agency) and CIFOR with the Federal Rural University of Pará.

CHAPTER 3

Creating Monitoring with Rubber Tappers in Acre, Brazil

Magna Cunha Dos Santos, Samantha Stone, and Marianne Schmink

I N 1996, A COMMUNITY FORESTRY MANAGEMENT project started among rubber tappers in the Agroextractivist Settlement Project of Porto Dias, in the State of Acre, Brazil (Figure 3-1). The Porto Dias forestry project focuses on the sustained production of timber using low-impact technologies to selectively log the forest. It is the second community-based timber management project in Brazil to be certified by the Forest Stewardship Council. There are many technical and social unknowns associated with the project, including its impacts on local people's well-being and rubber tapping as a livelihood, and about its ability to help preserve the area's forest cover and its future as a reserve.

To address these uncertainties, a program of adaptive collaborative management (ACM) started in Porto Dias in 2000. The three main players were the rubber tapper community of Porto Dias, *Centro dos Trabalhadores da Amazônia* (the Amazonian Workers Center, an NGO involved in rubber tapper issues in Acre), and a facilitator, a forestry specialist.

The ACM project set out to work with local communities, including Porto Dias, experimenting with the sustained production of timber as an alternative to more exploitive forms of land use, such as extensive agriculture and cattle ranching. Its core assumption is that communities' participation in (re)defining project impacts and related decision making strengthens local capacity to deal with future challenges. A critical part was therefore the construction of a participatory monitoring process based on criteria and indicators. Such monitoring would help in four ways: (1) to understand social, ecological, and economic impacts, (2) to improve communication between groups in the communities, (3) to reduce conflicts between groups, and (4) to facilitate collective learning and action.

This chapter describes the process of creating such a collaborative monitoring system in Porto Dias, based on CIFOR's criteria and indicators framework. After four years, many challenges remain to implementing collaborative monitoring,

Figure 3-1. *Porto Dias, State of Acre, Brazil*

particularly in ensuring a genuinely participatory process that contributes to local community empowerment.

Context

The Porto Dias project area is located in the municipality of Acrelândia, about 140 km from Rio Branco, the capital of Acre. Created in 1987, it covers 22,145 hectares of forest. Almost 90 percent of the land area in Acre remains under forest cover, much of which is inhabited by rubber tappers. In the 1970s, the sale of land for extensive cattle ranches triggered the political mobilization of rubber tappers. Their social movement became critical in securing land tenure and resource use alternatives that leave the forest intact. A major victory was the creation of two new tenure arrangements. Legally known as extractive reserves (administered by IBAMA, the Brazilian Institute of Environment and Natural Resources) and agro-extractivist settlement projects (administered by INCRA, the National Institute for Settlement and Agrarian Reform), they are federally owned lands to which local populations have usufruct rights to extract renewable forest products at ecologically sustainable levels. Nevertheless, the dropping price of rubber has threatened the livelihood base of these communities. This has led to an effort to diversify livelihoods through the production of other nontimber forest products and, more recently, among some rubber tapping communities, timber products.

In 2003, about 100 registered families and 20 illegal settled families were living in Porto Dias. The community is socially diverse, with two main groups: rubber tappers, who use family labor for collecting, processing, and marketing forest products (mainly rubber and nuts), growing annual crops (rice, beans, maize, and cassava) with virtually no external inputs, and raising livestock on a small scale; and farmers (known as *colonos*), who maintain larger planted areas and more livestock.

When the Porto Dias agroextractivist settlement project was created, most families were rubber tappers. However, since 1987, many have sold their usufruct rights to other families, most of whom are *colonos*. Other trends, including growing markets for wood and agricultural products, have accelerated the shift in land use. The absence of policies, institutional support, and political organization has resulted in poor dialogue between the two groups about the area's future as an agroextractivist settlement project. Conflicts have emerged as a result of the *colonos'* pushing for changing it to an agricultural colonization area.

In 1996, the Porto Dias community forestry project began, and in 2004, 10 families were participating. The project is small because of established economic, ecological, and technical criteria and because, since it is linked to the Rubber Tapper Association of Porto Dias, non-members cannot participate.

Timber management based on reduced-impact technologies is relatively new in Brazil, and forest resources are a source of tension. The resulting uncertainties that surround the project in Porto Dias (Box 3-1) created a need for a systematic and thorough learning process that would permit ongoing adaptation.

Two core assumptions underpin the ACM work in Porto Dias: that for management to be sustainable, basic agreements are required between interest groups; and that collaborative impact monitoring (social and economic) forms the basis for adaptive collective actions. Thus the Porto Dias ACM project sought to facilitate collaboration between different (sometimes conflicting) interest groups through explicit collective learning. By analyzing progress related to established objectives, continual adjustment would be more likely.

Box 3-1. Low-Impact Harvesting and Benefit Sharing in Porto Dias

The project applies harvesting techniques that minimize impact and maintain stocking levels and ecological niches. Harvest levels are low (about one to two trees per hectare every 30 years), and each family reserves a 300-hectare area of forest specifically for timber harvesting. To guarantee the 30-year cycle, each year only one 10-hectare plot is harvested per family. The annual harvested volume target was originally 1,000 cubic meters but in 2003 was reduced to approximately 350 cubic meters. About 12 species are harvested per year, based on ecological and economic criteria. Any profit is divided equally among the participants, even if harvesting was not entirely equal across the plots.

Collaborative Monitoring in Porto Dias

The project progressed in three phases: preparation, context analysis, and development of the actual collaborative monitoring system (Santos et al. 2001, 2003). The experiment with participatory monitoring started in 2000 with four organizations. CIFOR funded the first phase. The University of Florida designed the project, assembled the research team, organized the initial workshop, supervised the first phase of fieldwork (through Peter Cronkleton), provided financial support to the facilitator, and gave methodological assistance. Although the Amazonian Workers Center was the main contact organization, *Grupo de Pesquisa e Extensão em Sistemas Agroflorestais do Acre* (PESACRE, Group of Research and Extension in Agroforestry Systems of Acre, an NGO) supported the facilitator while she worked there. Table 3-1 details the activities undertaken during the first three years.

Phase 1. Introducing Collaborative Management

The preparation phase was critical for developing a mutual understanding among the possibly interested parties and for discussing some methodological options. Each party's level of interest had to be gauged. In May 2000, a first ACM workshop brought together key forestry actors in Acre: representatives and local leaders of three communities involved in community-based timber management projects (agroextractivist settlement projects in Porto Dias and Chico Mendes, and an agricultural settlement project in Peixoto), government agencies, NGOs, technical specialists, and researchers.

The workshop was followed by meetings (Table 3-1) to negotiate a formal collaboration between interested organizations and the communities. In the community-based meetings, discussions focused on understanding the level of local interest in using information to empower themselves in forestry management-related negotiations and activities. Some methodological options that had been presented at the initial workshop were discussed in more detail.

Phase 2. Participatory Context Analysis

Three communities were involved in this phase, Chico Mendes, Porto Dias, and Peixoto.[1] Field activities aimed at identifying the major issues and problems faced by the communities, such as impacts of the timber projects and tenure conflicts, from the perspective of the communities.

Diverse methods from CIFOR's methodological menu (Colfer et al. 1999a, 1999b) were used (Table 3-1 and Box 3-2), depending on the type of information being sought. Participatory methodologies, such as participatory mapping and future scenarios, were important for identifying social, economic, political, and ecological issues related to the management of natural resources that might be of interest for the community to monitor. These methods were also important for identifying important community actors and ways to involve more groups in the collaborative monitoring work. For example, the "who counts" matrix (Colfer et al. 1999b) deepened insights into the link between people and the

Table 3-1. *Phases of Monitoring for Adaptive Collaborative Management in Porto Dias, 2000–2004*

Phase	Contribution to Monitoring	Activities and Methods
1. *Introducing ACM and collaborative monitoring to communities* (May–July 2000)	• Increase understanding among partner organizations and community representatives about ACM, collaborative monitoring, and CIFOR methods. • Discuss usefulness of C&I framework for community-based forestry (timber) projects in Acre. • Discuss adaptation of participatory methods used by CIFOR. • Assess level of interest among possible partners and communities and their local associations. • Plan fieldwork for next phase.	1. *Workshop:* • Presentation of ACM proposal. • Group work (technical staff and communities) on vision of ideal forest. • Organizations presented work on forest management. • Discussions on idea of C&I. • Presentation of participatory methods (based on CIFOR's work), plus testing of some methods. 2. *Meetings with technical staff* in each partner organization. 3. *Community-level meetings:* Presentation and discussion of ACM proposal. 4. *Organization of fieldwork:* Discussions with organizations about context analysis process.
2. *Participatory context analysis* (July 2000–February 2001)	• Collect baseline data about community (interest groups, conflicts, socioeconomic aspects, etc.). • Identify community priorities for issues that require monitoring.	1. *Identification of interest groups:* Semistructured interviews, matrix of "who counts." 2. *Identification of key themes for monitoring:* See Phase 2 and Box 2. 3. *Feedback and discussion of community issues and concerns:* Community-conflict diagramming (Venn diagram), future scenario mapping to identify possible community-level changes. 4. *Writing of reports* based on data collected and discussions with communities.
3. *Construction of collaborative monitoring system (in Porto Dias only)* (February 2001–March 2002)	• Raise interest among all partners in participatory monitoring. • Clarify role of facilitator with technical staff and communities. • Construct simple C&I for impact monitoring. • Provide technical staff with better understanding of community processes and dynamics to improve community relations. • Build confidence among community members • Link technical, social, and economic aspects of timber project to enable internal evaluation of project.	1. *Open consultation about implementing participatory monitoring:* Community conflict maps, future scenario mapping to identify possible changes, and discussions with technical staff. 2. *Identification of strategies with potential partners (organizations and communities):* Meetings with technical staff in town and community meetings (using card and diagramming techniques). 3. *Detailed work plan:* Discussions between facilitator and town-based technical staff. 4. *Development of principles and C&I:* Initial filtering of CIFOR generic principles, criteria, and indicators through discussions with technical staff and with communities about issues and concerns raised during context study (using future scenarios, participatory mapping, semistructured interviews with managers and nonmanagers).

(continued)

Table 3-1. *Phases of Monitoring for Adaptive Collaborative Management in Porto Dias, 2000–2004 (continued)*

Phase	Contribution to Monitoring	Activities and Methods
		5. *Thematic workshops with communities.*
		6. *Pamphlet* to aid community and technical staff in recording data, using identified C&I.
4. *Implementation of system and ongoing adjustments* (April 2002–)	• Initiate monitoring system with communities and technical staff. • Ensure ongoing refinements to system.	1. *Monthly facilitation meetings* with Rubber Tappers Association to monitor timber project and other issues and problems in reserve (community meetings using participatory methods, including interest group matrix to identify conflicts, interests, and strategies).
		2. *Monthly meeting with rubber tappers* harvesting timber to evaluate and adapt technical and economic aspects of timber management project (group discussions).

forest in terms of their rights and power, dependency, and local knowledge. The diverse participatory methods used are summarized in Box 3-2. In addition to participatory methodologies, semistructured interviews with families and group interviews were used to elucidate the heterogeneity of land use and access to resources and benefits.

Based on the context analysis, five general concerns emerged in the Porto Dias project as possible foci for collaborative monitoring:

• social, economic, and environmental impacts of the timber management project (e.g., land use, tenure security, migration levels, distribution of (in)direct benefits);
• social organization and empowerment of different groups;

Box 3-2. Methodologies for Understanding Context and Concerns

• Participatory mapping: to analyze intergenerational security and access to resources and identify socioenvironmental conflicts.

• Future scenarios: to understand personal ambitions and plans, plus perceived landscape and resource use changes.

• Historical-ecological matrix: to understand how resource availability and access had changed over time, plus causal factors.

• Card sorting: to analyze rights and means for forest management by identifying interest groups with access to and benefits from the resources.

- efficiency of the management model (e.g., long-term viability of community control of costs, benefits, rules, and practices);
- internal and external conditions affecting project management (e.g., collaboration and communication mechanisms internally and with outside powers); and
- access to markets and basic services.

The purpose of these themes—the "key community concerns"—was to assess the extent to which the basic template of principles outlined by CIFOR (1999a) reflected the local reality. Participants compared these five themes with the principles in the generic template and decided to focus on only certain ones.

Phase 3. Constructing the Monitoring System

Developing strategies with partners. After the context analysis and community feedback sessions to verify information, the actual collaborative monitoring system needed to be developed. We held many meetings with partners to discuss initial results and identify how to proceed. The negotiations were complicated by differences, primarily among partner organizations, in political interests and technical perspectives. In the end, only Porto Dias and the Amazonian Workers Center decided to pursue a collaborative monitoring system. Therefore, Phases 3 and 4 took place only in Porto Dias.

Detailing a work plan. The monitoring work plan built on the existing interactions between the Amazonian Workers Center and the Rubber Tappers Association. The monthly association meetings and monthly meetings of the 10 families harvesting timber (as of 2004) were identified as the forums for elaborating a work plan. It had quickly become clear that differences between rubber tappers and *colonos* made it impossible to elaborate a single work plan and set of criteria and indicators for the entire community. Thus the decision was made to focus the ACM work on the rubber tappers. Nevertheless, to ensure that everyone's voice was heard, semistructured interviews with key informants and three group interviews were conducted (a sample of nonparticipating rubber tappers, participating rubber tappers, and *colonos*). Attempts were made to involve women in some of the interviews, but in most cases, their perspectives on the forest management system did not emerge clearly.

Discussing how to work together improved communication between the Amazonian Workers Center's technical staff and the community and clarified the interests of both. The center was interested in defining criteria and indicators to continually improve the project and respond to wood certifiers' prerequisites and conditions. Community members, specifically the rubber tappers, were interested in improving people's well-being. Creating a viable and collaborative work plan that would bridge these interests required that participants reach a consensus on the focus of the timber project. Four critical issues were identified:

- formulating participatory conflict resolution mechanisms;
- establishing criteria to select new participants and define clear and dynamic processes for consultations with groups affected by the forestry management;

- improving technical and administrative components of the timber management project; and
- measuring social impacts of the project.

Identifying principles, criteria, and indicators. The next step in the process built on CIFOR's framework (Ritchie et al. 2000) and entailed discussions with the Amazonian Workers Center and the community (see Table 3-1 for a list of methods used). At the community level, four questions guided discussions on principles and were followed by more detailed questions (Box 3-3):

- What is your dream of the forest in the future?
- What key elements are needed for sustainable forestry management (in order of importance)?
- How can you recognize a sustainably managed forest?
- How can you recognize good community management?

The questions helped make the principles from the initial generic set (CIFOR 1999a) more appropriate for Porto Dias and were critical in identifying areas of consensus and divergence.

Identifying, discussing, filtering, and systematizing the criteria and indicators took about six months before an initial set emerged. Three principles formed the heart of the monitoring framework:

- Human well-being is a right of all in the Porto Dias agroextractivist settlement project.
- The external environment (e.g., policies, partnerships with external institutions) is favorable for community forestry management.

Box 3-3. Specific Questions Discussed with Rubber Tappers in Porto Dias

1. The project is six years old. Were the initial uncertainties the same as those you face today?
2. What are the indirect and direct benefits generated by the project?
3. According to the nonparticipants who are not receiving direct benefits, does the project have any negative effects?
4. Did the project change community life in any way? If so, how?
5. What indicates whether the project has changed community life in any way?
6. How can you measure whether the project is being adapted?
7. To what extent did the project improve community-level communication?
8. To what extent did the project maintain the health of the forest?
9. What else needs to be improved?
10. How will the income improve life in the community?
11. What will enable the long-term sustainability of the project in Porto Dias?

- Forestry management improves the living conditions of people in the Porto Dias agroextractivist settlement project.

As per the CIFOR methodology, criteria and related indicators were identified for each of these principles.

Conducting thematic workshops. After constructing a basic set of criteria and indicators, three thematic workshops were held with the communities to discuss in more depth the monitoring concepts and mechanisms. At the first workshop, the suggested criteria and indicators were verified and validated. The community continued to have difficulty in understanding the concept of monitoring, and therefore more in-depth discussions focused on the purpose of monitoring and on linking data collection and analysis. The other two workshops led to more detailed planning of the monitoring system, based on local capacities and on existing informal and formal monitoring activities.

Writing a community monitoring guide. One of the main challenges was to ensure that community members participate in the monitoring system. A document outlining the agreed monitoring process, including information registration, was written. For each indicator, a participatory method for monitoring and verifying any changes was specified. The community allocated roles for each task and frequencies were identified. It was decided that analysis of the information collected would take place during the monthly meetings, when project decisions and strategies are usually defined.

Phase 4. Implementation

In 2003, monitoring and registration of information began but in 2004 came to a virtual standstill. Although registration methods were continually being refined to facilitate use, several difficulties emerged. For example, significant events were observed by community members but could not be related to the indicators. Also, community members could not reach a consensus on whether to appoint two people to take on core monitoring responsibilities. In addition to methodological and administrative difficulties in implementing the monitoring system, conflicts in the agroextractivist settlement project also had an effect. Most notably, its *colono* residents decided in 2004 to implement their own, separate timber management project, which many rubber tappers considered ecologically unsustainable. This heightened the conflict between the two groups and diverted attention and interest away from the monitoring system as a means to foster collective discussions and a shared sense of responsibility for timber management in the reserve.

As a result of these problems and other related difficulties, in 2004, the Amazonian Workers Center decided to rethink its monitoring program in Porto Dias. Since dynamic social changes had occurred, the center decided to interview families again to update the "key community concerns" and the foci for monitoring. The center also decided to hire a consultant to help simplify the monitoring system and identify participatory methodologies for collecting, registering, and discussing information that would be more accessible and compelling to the community.

Lessons

Conceptual and Methodological Challenges

Although participatory processes of developing collaborative monitoring were effective in involving the community, community members had difficulty understanding and relating to the hierarchy of elements as formulated in CIFOR's criteria and indicators framework (1999a). Local leaders, having been involved in most of the discussions about ACM and other projects in the reserve and being generally more educated, dominated the discussions and wielded more power in the community as a result of their formal roles in the association. To be effective at the community level, the language of criteria and indicators needs to be clearer and based on locally relevant examples (Chapter 2).

Creating an understanding of collaborative monitoring and constructing the actual system demands a heavy time investment on the part of the community. This comes on top of ongoing community decision-making activities. Frequent, long meetings and activities made monitoring-focused discussions impossible at times. To ease the burden of monitoring and ensure a good fit with the rhythm of the community, we allowed ample time. This resulted in a slow and sometimes fragmented process.

The monitoring of environmental indicators has proven particularly difficult, dependent as it is on technical expertise. Funding agencies and wood certifiers demand such information, yet NGOs, such as the Amazonian Workers Center, have limitations. The lack of staff with ecological monitoring expertise has restricted that aspect of the work. Community indicators were identified (but not yet used) to assist in monitoring environmental impacts of the project, but more scientific indicators will be needed to satisfy funding agencies and certifiers.[2]

A particular challenge is how to use problems that have arisen in the community (such as dependency on timber and the abandonment of other production systems) in a constructive way to enhance collective learning and adaptation, without generating conflicts and bitter disputes within the community and between the community and the Amazonian Workers Center. The timber project in Porto Dias has a long history of mistakes that have generated important lessons and triggered technical and administrative improvements.

A related challenge is discouraging actions and behaviors that jeopardize the project. Learning does not necessarily imply positive collective action—it can equally result in detrimental decisions. For example, the Amazonian Workers Center and the families harvesting timber spent several cycles of the timber project discussing marketing problems, each year reflecting on lessons learned from the past year's experience to find a better way to sell their timber: that constitutes positive collective learning and action. However, some families, realizing how difficult it was to market timber and being unsatisfied with dividing the profits, decided to pull out of the project altogether. From the perspective of the remaining participants and the Amazonian Workers Center, this example of learning was harmful to the collective effort.

Relational Challenges

Collaborative monitoring, by definition, requires clarity of roles and responsibilities for all partners. It remains a challenge to interest other agencies in assisting the community and the Amazonian Workers Center in monitoring. This includes government agencies that are fundamental for ensuring the success of the Porto Dias timber project through their capacity as auditors of the timber management plans and enforcers against illegal deforestation and settlement activities. A first important step is convincing the diversity of actors of the value of such monitoring and then ensuring clear linkages between them so that learning is genuinely shared.

Perhaps the greatest challenge is creating a space of trust and learning so that results can feed into improved actions. In Porto Dias, much time was invested in aligning work schedules and establishing priority themes for the monitoring. A facilitator was essential for mediating conflicts and differences. The facilitator, by virtue of not being affiliated with the Amazonian Workers Center, was seen by the community members as neutral and won their trust—and that was one of the most significant results at the community level. (The facilitator has since joined the center's technical staff and is now seen as a member.) Trust is also needed between community members and the Amazonian Workers Center's technical staff, who face time pressures from funding agencies and may be skeptical about collaborative monitoring.

Partnerships are, of course, not static. In Porto Dias, the ACM work started with many organizations and then focused on the Amazonian Workers Center, PESACRE, and the Rubber Tappers Association. Starting small in a socially complex process such as collaborative monitoring has some virtues. However, the involvement of other actors, especially government agencies, is vital. Public policies and commodity markets greatly affect production issues, such as species choice and prices, but communities have limited influence over these external forces. Even if these were monitored, intervening in policy decisions would be difficult. The Amazonian Workers Center is currently working to increase dialogue and joint actions with state government agencies and thus clarify its role (and its limits) in the Porto Dias agroextractivist settlement project.

The great diversity of community interests, inequalities, conflicts, and perspectives on forestry management affected the learning process in several ways. In the absence of social institutions that could mediate conflicts and make possible genuinely collective action, the facilitator was critical. Porto Dias is a divided community with a history of many conflicts, and facilitation required much patience, skill, and diplomacy. Opportunities for dialogue had to be created and new relationships of trust forged—both of which take time. One group could not be privileged over another, but universal participation simply was not feasible. Building capacity and strengthening community organizations, specifically of the Rubber Tappers

Association, was accomplished through discussions of the agroextractivist settlement project's natural resource utilization plan and definition of strategies, such as the development of rules and regulations to strengthen the association's capacity to enforce the natural resource plan in the reserve.

Despite the presence of a dedicated facilitator to link the groups, decisions and debates, misunderstandings, conflicts, and incorrect expectations were unavoidable, both between the Amazonian Workers Center and the community and within the community. Not all planning processes involved the community. Different perspectives among the technical staff and the power structures made collective decision making involving the Amazonian Workers Center and the community difficult. The center's next efforts will focus more on ensuring collective planning, which will help establish a common language and ensure transparency in decision-making processes, thus preventing poorly planned technical interventions.

Collaborative monitoring has risks and limitations that communities and all other partners must understand. As a mechanism to enhance understanding and decision making, it has great potential. However, if there is no time for communities to acquire the necessary skills and insights, it loses its local value and becomes yet another externally driven evaluation tool.

A significant unknown concerns the sustainability of monitoring systems. Communities need time to appreciate and institutionalize the action-learning cycle, including monitoring. In the absence of a skilled facilitator, who can ensure that monitoring is kept alive? Thus far, efforts to sustain critical reflection and adaptive management by the diverse interests in Porto Dias are inconclusive. One issue is clear, however: investing in better social organization and relations is a priority, since existing resource-related conflicts threaten to erode the social unity needed for collaborative learning.

Since 2003, efforts to extend collaborative monitoring in Acre have grown. Current national and state forestry policies favor the focus on information in support of forestry management, and an organizational consortium funded by the U.S. Agency for International Development has provided the financial basis for a larger-scale endeavor in Acre. This should give the communities, governmental agencies, and NGOs more time to understand some of the challenges to forest management in Porto Dias.

Notes

1. Two other government agencies were involved, providing a point of contact in the communities and assisting with field logistics: EMBRAPA, the national agricultural research service, in Peixoto and SEFE, the Executive Secretariat of Forests and Extractivism, in Chico Mendes.

2. The Amazonian Workers Center has written a proposal for more technical environmental monitoring.

PART III

Building on Existing Monitoring Systems

CHAPTER 4

Imposing Indicators or Co-Creating Meanings in Nepal

Krishna P. Paudel and Hemant Ojha

O VER THE PAST TWO DECADES, **the** government of Nepal has handed over patches of forest to local communities. These communities form forest user groups as part of the nationwide community forestry program supported by the Master Plan for the Forestry Sector (1988), the Forest Act (1993), and the Forest Rules (1995). The main aim of this program is to provide a legal basis for access to forests by local communities as a means to improving livelihoods, as well as to reverse the trends of forest degradation and deforestation, particularly in the middle hills of the country.

Recent studies have demonstrated that the community forestry program has in many instances led to improved forest conditions. Nevertheless, some studies have highlighted problems, such as underutilization of forests and poor people's low level of access to the benefits from community forests.

Community forestry implementation suffers from two main difficulties that have contributed to these negative impacts. First, within forest user groups, decision making has been dominated by a minority of community members who very often depend least on common property forest resources. Field experience has also shown that the district forest office and range post staff often drive decisions through the elite members of the forest user group, even including the decision to form the group. Second, other organizations involved in community forestry face difficulties in identifying support services and skills that are appropriate and responsive to the needs of forest user groups and specific groups within them, and accordingly in understanding the impacts of their current services and policies. A related concern is inadequate communication between stakeholders.

The problems highlight a need to enhance the ability of forest users to undertake two crucial steps. They need to plan systematically for forest management on the basis of needs analyses and forest resource–related social, institutional, and economic factors. They also need to demand services from and provide feedback to other organizations, including the field offices of the Department of Forests.

This arena was the scene for our research on monitoring, which we initially thought was the missing piece that would enhance forest user groups' ability to undertake effective forest management (see also Pokharel and Grosen 2000). From November 2000 to May 2001, we carried out intensive field research to develop a monitoring process and indicators aimed at supporting forest management in five sites (four of which were forest user group sites) in the Kushmisera Range Post of Baglung district in West Nepal (Figure 4-1).

Original and Revised Research Intentions

The original research purpose was to assess stakeholders' perceptions or expectations of common property forests and, with those insights, to collectively identify indicators that would support local forest management. Our research was based on two assumptions. First, we assumed that local communities were firmly in control of their community forest resources and would undertake the identification of forests and their users and preparation of operational (management) plans as specified in the community forestry guidelines. Second, we assumed that community forest users were aware of the government community forestry policy objectives, and that all the aspects of forest management were considered in the forest user group's operational plan.

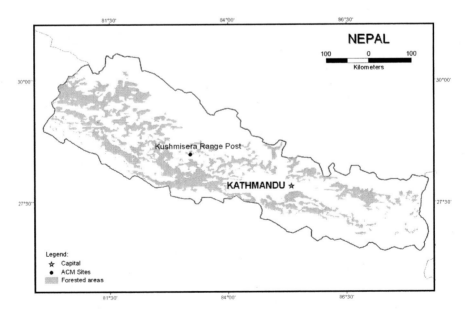

Figure 4-1. *Kushmisera Range Post, Baglung District, Nepal*

Based on these assumptions, the research project originally proposed that a set of criteria and indicators be developed that would help assess whether a community forest was managed as specified in the operational plan and that the process used to arrive at such criteria and indicators could be documented. Explicit monitoring systems at the level of local forest management were considered necessary for two reasons: to enhance the internal action and learning processes in local forest management institutions (in particular, the forest user groups) in pursuit of livelihood security and sustainable resource usage; and to raise local people's interest in stakeholder interaction through active monitoring, to improve the relevance of and accountability of interventions, service delivery, and policy formulation.

In line with the original research purpose, the research team invested much initial effort in finding indicators to monitor. But we soon started to doubt this focus.

Our research began with a pilot phase using a participatory action research approach in Pallo Pakha Forest User Group. We developed a general understanding about the forest users through meetings with key representatives, the group committee, and subsequently in *toles* (hamlets, generally 10 to 20 houses and the level at which informal communication usually takes place), with as many households as possible. From this, we asked each *tole* to select representatives who would work with us on the detailed investigations in the next phase.

At the first representatives' workshop, after explaining the original research purpose, we started discussing indicators. Two difficulties emerged immediately that helped shape our strategies with other forest user groups and led us to conduct a second workshop.

First, we found it difficult to convey the meaning of monitoring and indicators. Translated literally, monitoring and indicators are *anugaman* and *suchak*, Sanskrit words used mainly by Brahmin groups, with other community members relying on their interpretations. *Anugaman* is also viewed rather negatively by forest guards and local village elite, since senior officials from the national, regional, and district headquarters often come to assess their work. Even approximate local words or phrases relating to monitoring or investigation have negative connotations or are perceived as activities conducted only by officials and technicians. Phrases such as *rekh dekh* ("keeping an eye on something"), *lekhajokha* ("weighing up and writing"), and *khojbin* or *chhanbin* ("investigation," usually referring to investigating someone else) were considered but did not help develop a common understanding about the monitoring process.

Second, we had difficulty developing consensus about the meaning of specific indicators. The different interests of workshop participants clearly shaped their perceptions on indicators. While discussing "improved forest condition," for example, a member of the group who had rented land for grass cutting (on which her cattle graze) said the indicator would be "good ground cover with grass"; the forest guard had as his indicator "forest cover where we can't see people moving inside the forest."

More significant perhaps was our observation from these initial meetings and workshops that an undue focus on indicators to assess planned activities took attention away from more important prerequisites that would enable active management by the forest user groups. For example, we identified an imbalance in the representation of individual members and interest groups, as well as in the opportunities for forest user groups to generate new knowledge through their forest management activities. Some interest groups appeared to have minimal or no representation in the annual planning meeting where important decisions are made. These meetings are mainly attended by range post staff and members of the forest user group committee, which is the executive body elected by group members. We also noticed that the majority of the members, including those belonging to the most active or strongest forest user group, were not aware of the government's community forestry policy objectives. Apart from a few forest user group committee officials, mainly the chairmen and secretaries, very few people knew about the existence of operational plans. Some people did not even know that they had a community forest. In such a situation, it would have made little sense to talk exclusively about monitoring indicators.

Such structural imbalances would certainly affect the quality of any monitoring-related adaptive management, so we clearly needed a process to identify indicators (and other aspects of a monitoring process) that would reflect the variety of perceptions and interests among forest users. In this way, indicators would enable each interest group to analyze trends in the forest after establishing a baseline of information.

Hence, our two basic assumptions were not applicable in these forest user groups at the time, and our indicator-focused work had to be revised. The project's purpose was adapted to allow space to develop and assess participatory action and learning approaches in general for common pool resources management and biodiversity for sustaining livelihoods. This included monitoring systems—not just indicators—that support stakeholders in forest management. Therefore, we let go of the narrow research focus on monitoring systems driven by indicators.

Assessment of Issues

With this revised research purpose, the research team facilitated the next steps of the field investigation after forming a wider research group that included forest user group members.[1] A group inquiry approach was used to bring together and analyze the issues from the *toles,* develop a set of proposals for short-term and long-term solutions, and set clear objectives for each proposal. These proposals were then discussed by the *tole* representatives with the other households in their *toles.*

All the household members of all the *toles* then came together in an assembly to decide on the priority issues, develop a work plan for each decision, and plan ways to monitor both implementation and impact. Where no consensus on particular issues was reached or further research was needed, the information requirements and research methods were also discussed. After this, the forest user groups started implementing decisions and monitoring progress. Implementation and monitoring began just as our research project was ending; hence our observations are only partial and preliminary.

The core methodology that drove the monitoring fieldwork with community members from the five sites was participatory action research. We developed case studies to examine local monitoring practices within the forest management planning framework. Our field investigations focused on methods that included different people's needs and interests, and that enabled group members to use monitoring to learn from new experiences. In the beginning of the process, we identified major issues and challenges of participation by the various interest groups, which we then used to set up basic norms to follow during meetings and workshops. One such challenge was the representation of women and minority subgroups, such as occupational castes and landless people. The range post staff members were involved throughout so that they could develop capacities to continue adapting the process in the future.

Throughout the research, the core research team, including *tole* representatives, constantly assessed and reflected on the usefulness of the methods to make continual adaptation of the research methodology possible. Various methods were used, including resource and social mapping, village and forest walks, situation analysis (via a web diagram), visioning, wealth rankings, focus group discussions, forest resource assessment, and sample plots. Several methods were used initially for understanding the situation and then for monitoring. For example, wealth ranking in Bhane Forest User Group was used by *tole* representatives during one workshop to identify the various categories of forest product users. Subsequently, it was used to see how different wealth categories were represented in the forest user group committee and among *tole* representatives. After the workshop, the representatives revisited the wealth ranking in various *toles* to see whether the categorization of households needed revision.

Finally, at the end of the initial research project, together with the *tole* representatives, we analyzed the research process and planned subsequent activities, such as implementation of planned action and research followup. This involved reflection on the methods, tools, and techniques, the sequence in which these were used, the role of the facilitators, and possible adaptations for the future activities of forest user groups, as well as range post staff and the Federation of Community Forest Users.

We developed a one-year followup plan to the initial research project to gain more insights into the sustainability of the process at the forest user group and range post levels and the constraints encountered by the forest users and range post staff alike. The research followup involved two groups besides the external researchers. For each forest user group, a team of three members representing the forest users and the *tole* was formed. Similarly, a team was formed at range post level consisting of range post staff members who had been involved in the process, including members of the Federation of Community Forest Users network. In this chapter, however, we report only on the initial research.

Methodological Results

The research led us to identify a generic and simple process (Figure 4-2) that can enable forest users in different locations in the hills to identify priority issues relating to forests and livelihoods, undertake critical inquiry and self-analysis, negotiate interests (especially of disadvantaged groups), and develop flexible plans that allow for experimentation, monitoring, and review. The research team also prepared a set of guidelines for information requirements and data collection methods to assist facilitating organizations that undertake this process with forest user groups in the future.

The core elements of the methodology are straightforward and replicable, making clear how the different groups contribute to Adaptive Collaborative Management (ACM). After developing the system, the forest user group members will be able to use the process on their own and adapt it to their circumstances. Although outside facilitators are needed to help initiate the process, their involvement can taper off.

Because forest user group members as well as range post staff were involved in all stages of the research activities, we were able to develop a practical process for monitoring that moved away from the limited notion of indicators and therefore had local relevance. A followup study (Paudel et al. 2003) conducted in early 2003 indicated a strong feeling of ownership among the community members and range post staff over the monitoring system and the process of developing it. The study showed that community groups were expanding the process to other aspects of community development, such as school management.

The methodology that we developed through our participatory action research process is different from the existing approach in Nepal for supporting forest user groups in several ways.

First, it reaches the majority of forest user group members, beyond the committee officials, by actively engaging the *tole* level, thus making it more likely that all interest groups' views and concerns are considered. In the past, the committees had been assumed to represent the forest user groups in all their diversity. This occurred

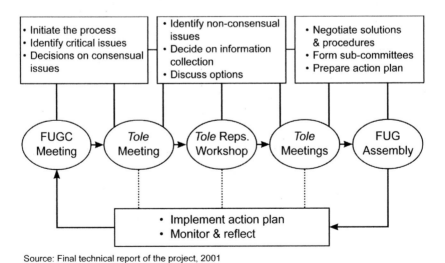

Source: Final technical report of the project, 2001

Figure 4-2. *Forest User Group Participatory Action and Learning Cycle*

Note: circles = core activities; boxes = purpose of each activity.

because committee membership was based on interest groups without any minimum quota per interest group, and few members from certain interest groups (particularly minority occupational castes and women) were selected. A provision for women's representation was ignored by some groups. Moreover, despite the provision for *tole* meetings in the official process, there were no functional meetings. This passive representation paralyzed functioning at the *tole* level: members in the *tole* became disengaged because they assumed that once a representative had been chosen, it was his or her responsibility to represent all interest groups in the *tole;* and the representative had little sense of responsibility because he or she was selected (not elected) and hence had no functional relation to the rest of the *tole.* In our action research process, we emphasized the importance of involving all forest user group members (not just the representatives), and that reinvigorated the groups. Active social relations exist at the *tole* level, where effective communication, interpersonal sharing, and exchange happen—and not at the forest user group level. So giving the *tole* meetings a prominent role opened the door for awareness and more critical reflection on the community forestry process. At *tole* meetings, issues, conflicts, and points for negotiation can be made visible.

Second, our approach provides a sequential framework for information collection and analysis, explaining who is required to do what and the objectives for each step or activity. As mentioned above, initially there were no *tole* meetings for planning and reflection. The forest user group committee simply collected information as asked and required by range post and other development project staff.

Because considerable confusion can occur in collaborative forms of resource management, clarity of roles is critical. Our process explains the stage at which the interests of various interest groups and individuals are best brought together for discussion and negotiation, and the ways in which this can be done (Malla et al. 2002). As part of this, we stipulate the need to promote a deliberate and practical strategy of communication between forest user group and other local stakeholders, including the range post and district forest office. This is effective only if the existing asymmetrical power relations among the groups within a forest user group, as well as between the group and government staff, are tackled. We did this by promoting forums and platforms for the open exchange of views and arguments.

Most important, the ACM-inspired version of community forestry management that our research project helped define also incorporates mechanisms to identify and pinpoint uncertainties, knowledge gaps, and practical strategies for more deliberate observation, experimentation, and analysis for learning. For example, throughout the research project, we emphasized local research teams so that we, as outside researchers, could hand over the research process to these teams. Thus we were able to be process facilitators, as well as observers of the process as it evolved.

Lessons

A narrow focus on indicators, as defined by the original research purpose, left little time for us, as researchers, to understand the processes through which learning takes place within forest user groups, and to which the targeted indicators—which are, after all, only part of a learning process—could perhaps be linked. For example, processes for enhancing participation of various interest groups, developing a common understanding of meaning, and defining the purpose of forest management are all crucial for effective and equitable management processes. Only with a broader understanding of existing learning and participatory reflection processes would we be able to identify useful and cost-effective indicators and understand the contribution that indicators—only one of many learning tools—could make locally.

Another important observation from our study concerns the role of external facilitators in collaborative learning processes. External facilitators can make several significant contributions:

- suggesting the overall research and communication process in terms of a sequence of meetings;
- assisting clients in bringing together different issues, interests, and perceptions around the major issues of group functioning and the resource itself, particularly where these conflict with each other;
- helping negotiate solutions; and
- providing a broader picture of social, economic, environmental, and political realities and thus developing a common basis for making transparent decisions about common property resources (in particular, regarding equity, power relations, and environmental degradation) and helping participants relate this to their own group performance.

In such roles, facilitators must minimize impositions and engage in open and empowering dialogues that enable local actors to find and articulate their own ideas and options. This requires a philosophical commitment to a participatory process of change and demands considerable social skills in creating learning groups (cf. Groot and Maarleveld 2000).

Therefore, our key lesson is that an externally construed, indicator-oriented monitoring process does not necessarily help the learning and planning process of local actors or those operating at higher spheres of decision making to cope with their situations. As soon as we understood this, we were able to step back from focusing on indicator development and begin finding a strategy for a collaborative process of learning, negotiation, communication, and improved management of forests. Monitoring remained at the center, not as a rigid and hierarchical framework of indicators, but as a dynamic link between learning and planning.

Generally, when development activities are facilitated by outsiders—be they government officials or donor-funded project staff—fieldworkers often impose on client organizations either the notion of indicators or the actual indicators. These impositions are neither feasible nor useful for either party. Our study suggests that it is unrealistic to develop a systematic and hierarchical monitoring system that combines indicators from all stakeholder groups while ensuring each interest group's active participation. In our case, monitoring ceased to be a separate activity. Instead, viewing monitoring as the reflective process helped the group members critically review their plans of action, outputs (as well as outcomes as the consequences of the process), and related problems. Individual as well as group interests were reflected during the discussions and negotiated through the process of collective (group) analysis. This was driven by the process of reorienting power relations among people within and outside the forest user groups.

Acknowledgments

This chapter is based on research carried out by the International and Rural Development Department of the University of Reading and Forest Action, a national nongovernmental organization, and the Nepal-UK Community Forestry Project, now Livelihood and Forestry Programme, funded by the (UK) Department for International Development through its Natural Resources System Programme. We acknowledge the conceptual guidance of Dr. Yam Malla and other team members in the research.

Notes

1. The research group differs from the executive committee in that it is larger and its membership is supposed to change annually. Furthermore, the research group considers the committee a stakeholder within the forest user group.

CHAPTER 5

Helping Village Stakeholders Monitor Forest Benefits in Bolivia

Peter Cronkleton, Robert E. Keating III, and Kristen Evans

A CRUCIAL STEP FOR COMMUNITIES LEARNING TO maintain functional forest management plans is the definition and use of collective
monitoring to generate basic information for decision making. This case
explores our experiences assisting two Guarayo indigenous villages in Bolivia
to implement community forestry management projects (Figure 5-1). It illustrates how we worked with these communities to develop simple systems to
monitor their projects' distribution of benefits. The systems focused on the
collection, presentation, and discussion of information that would allow the
villagers to control wage payments to residents who had invested labor in the
project. Although there is still room for improvement in the system, village
residents are now able to track the flow of benefits transparently and have
avoided major conflicts.

In Bolivia, commercial timber management is a new activity for indigenous communities. They must learn a great deal to be able to form operating
plans that meet government-mandated sustainability guidelines. Members of
indigenous communities normally monitor their environment and production
systems in a variety of ways, sometimes informally. Introduced projects often
require new monitoring activities, and for communal projects, these monitoring
activities must entail some collective evaluation. It is a challenge to introduce
monitoring systems that are locally relevant and reflect local capacity. Outsiders often have different perceptions of what needs to be monitored and how it
should be done. By working together with local stakeholders, it is possible to
develop monitoring systems that work and are used by local decision makers. In
the Guarayos cases, we focused local monitoring efforts on the benefit distribution system because it was poorly understood by community members and held
much potential for conflict.

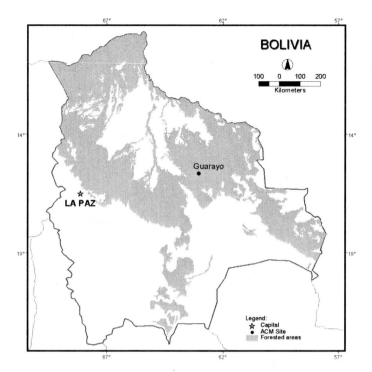

Figure 5-1. *Guarayos Indigenous Territory, Department of Santa Cruz, Bolivia*

The Challenges of Monitoring Wage Benefits

The neighboring villages of Cururú and Salvatierra in the Guarayos indigenous territory each initiated a community forestry project with assistance from CIFOR and the Bolivian Sustainable Forestry Project (BOLFOR). In 2000, Cururú defined a 26,421-hectare management area near the village. A year later, the villagers of Salvatierra designated 40,886 hectares of surrounding forest for their management plan. Once underway, village management organizations would marshal labor to carry out annual timber censuses, negotiate sales with timber companies, and control the actions of loggers on their land. These Guarayo forest management projects were designed to generate income through the sale of timber.

The management areas are in communal forests, so all village residents have the right to receive benefits from the logging. The villages' management strategies provide two ways for residents to benefit. They can invest labor in the plan and earn wages from the community's management organization once timber is sold. Alternatively, those not interested in working in the forest can benefit from the distribution of profits remaining at the end of the season after all costs have been paid. These profits are turned over to the entire community, to be invested in common needs. This dual approach to benefit distribution ensures that all have access to some benefits but provides greater rewards to those who help generate those benefits.

Although apparently simple, the strategy held unseen challenges because the village management organizations were new. Leaders lacked experience in managing large sums of money and carrying out the transactions necessary to pay wages. The situation was aggravated by the long periods between forest management activities and timber sales and the subsequent payments to individuals. The entire system would break down if the group was unable to develop a system to distribute wages to those who had worked.

This meant that over several months the villages would need to track who worked and how long so that their organization could pay them later. The task was further complicated by the timber companies' installment payments, which would prevent the village organizations from paying their members' wages completely at one time. Instead, they would need to track balances owed to each member to avoid misunderstandings or conflicts. The communities therefore needed monitoring systems that would be transparent and easy to understand, allow participants as well as nonparticipants to know what was happening, and provide the information to make decisions.

Ideally, the mechanism for distributing and monitoring benefits needed to be developed and agreed to before any income was generated. Creating such a system after the project was fully operational and actual money was in play would increase the potential for misunderstandings and conflict. However, it was difficult to initiate a collaborative monitoring system for benefit distribution when the timber harvests had not started and the organizations had no income to distribute.

We faced a challenge: how to start working on monitoring skills and systems when residents had little patience for discussing monitoring in theory and no benefit for them was apparent. Therefore, as an example we decided to use existing records they had kept to receive startup capital from BOLFOR.

Initial Monitoring

At the time that the BOLFOR collaboration started, there had been no income for locals to monitor. However, together we could examine records that village leaders had submitted to justify the use of startup capital from BOLFOR. Both communities had received financial assistance while they were developing their timber management plans. Most of these funds were used to pay a portion of daily wages earned by members working on the project—B$10 (about US$1.30) out of the total B$25 (about US$3.25) per day (November 2006 conversion rates)—with the remainder to be paid by the community management organization once timber was sold. BOLFOR's administration required that community leaders keep records of these transactions for funds to be released. The forms were an obligatory chore and leaders did not really conceptualize this as monitoring.

Members of the community organization knew that they had received this support from BOLFOR and that the leaders had recorded their names, how much they had worked, and how much they had been paid, but they were not aware of how or why this information was useful. By facilitating a discussion of these records, we hoped to show residents how they could track what they and others were still owed. This would allow everyone to know what to expect once the community had funds to distribute after the first timber sales.

Since the community had maintained these records—albeit to meet externally imposed obligations—they provided an excellent opportunity to discuss how such information should be collected in the future to monitor payment of the remaining portion of the wages. Rather than using an abstract example, we were discussing a case directly affecting the residents. Wages and benefits were the prime motivation for the management plan. It was crucial that information be recorded systematically in the future and disseminated to all community members to promote transparency and understanding.

Starting Out in Salvatierra

Our first attempt to generate interest in a wage monitoring system took place in Salvatierra in September 2001. At that point, the management organization had just finished its first census of commercial timber species in a 900-hectare harvest unit. The job had taken 26 days, and 34 individuals had worked for at least part of that period. The organization had used BOLFOR funds to pay B$7,890 (US$1,023) to members but still owed B$12,015 (US$1,564) that would have to come from the proceeds of timber sales.

We helped the leaders compile information from their records to be presented to the community during a general meeting. We transferred information from the labor records onto a wall-sized chart that listed each member's name and his or her participation in each activity. This included the total number of days worked, the total wages earned, the advance received from BOLFOR, and the remainder to be paid after the first timber sale. We also prepared a handout for each individual summarizing his or her personal information from the chart. During presentations, in addition to discussing the individual data, we pointed out the total days invested in the activity and discussed the importance of such information for future planning.

A major topic of discussion was the variation in labor investment by different members and the resulting variation in earned wages. For example, the census and inventory coordinator had participated in all forest activities and, as a result, had worked the most and had earned the highest wage. Evaluating this variation was important because if the group had not recorded this information, or if none of the residents had been aware of this variation, it could have damaged group cohesion. When it would finally be time to distribute funds to pay the wages, if residents did not understand that the difference in payments was due to variation in days worked, those paid less could perceive the system as unfair and become disgruntled. If the community ignored the variation in labor and divided the income equally, those who invested more labor could become disillusioned and less motivated to invest time in the future.

An interesting exchange occurred when one resident realized that he had not been credited with all his days worked. His first reaction was to shout that he was being cheated. After calming him down, the group checked the original records and learned there had been a mistake. He had worked eight days, but when the coordinator transferred information from a field notebook to the BOLFOR form, the eight had been recorded as a six. It was an honest mistake that taught everyone the importance of the records and of double-checking them.

Moving On to Cururú

After the first experience in Salvatierra, we repeated the exercise in Cururú. We reviewed the process a second time in spring 2002, before harvests began, to be sure that everyone understood. Once the timber sales started, the community would be monitoring its own money and paying the amounts owed to members. Without a system in place, this would have been difficult, especially since some residents were still owed for labor that they had invested two years earlier. Also, because logging companies paid in installments, the organization could make only intermittent payments to its members. As a result, leaders had to pay wages in installments, which required holding several meetings and maintaining records to track transactions.

In both communities, residents found the system useful and continued to record the same information for monitoring labor investments even though they no longer received funds from BOLFOR and were under no obligation. One reason was because the system worked. Residents who had been anxious about their wages appreciated knowing how payments would occur and were confident that they would get their fair share.

Evolvution of the Monitoring Systems

Although both groups adopted the labor-wage monitoring system, over time they began to diverge. Cururú continued to monitor actively, but Salvatierra's monitoring became more haphazard and less transparent. To a certain extent the difference was due to the idiosyncrasies of the two communities and their leaders, but nevertheless the difference provides significant illustrative lessons.

In Cururú, the public distribution of payments became the norm, and community leaders continued to devote time to preparing records to demonstrate the transactions in the community. Why? The considerable delay between the initiation of their project and the first timber sale had made residents very demanding. People complained about the lack of payment and some accused the leaders of holding money back. The leaders felt pressure to provide as much information as possible. They found that community members could easily understand the use of visual aids and personalized printouts. Also, the group faced a crisis early on that proved the importance of monitoring the distribution transactions.

The crisis arose over a misunderstanding about credit given to some community members and backed by the forest management plan. Before the first timber sale, community leaders reached an agreement with a local merchant to extend a small amount of credit to participating families. In the agreement, each family that had worked could spend up to B$250 (US$33). However, the debts had to be paid within a month or the merchant would charge the community interest. Because of miscommunication (and the merchant's interest in extending credit to make more sales, in expectation that the buyers would soon have cash from the timber sale), several members exceeded the credit limit and even exceeded the amount they would earn in wages. This began to generate internal conflict, since it was not clear how the community could ensure that all members would pay off their debts. It would not be fair to those who did not take credit to have to wait while the collective debt to the merchant was repaid. Many community members feared that everyone would have to pay for an action that had benefited only a few.

To avoid the potential conflict, they decided to increase the information included in the monitoring report for the first wage payments, scheduled to take place later that month. They created another column on the wall chart that listed each member's debt to the merchant. Those exceeding the limit had their debt listed in bold red letters. When the wage installments were paid, the treasurer first discounted the member's debt. Those who had not taken credit received their full payment, while those with debt found that much of their pay was discounted. This ensured that members paid the merchant without penalizing others. Those who had exceeded the limit had the added shame of being singled out in front of the community (one member even made a public apology to the rest of the group for having "taken advantage of their trust"). Although this may have been uncomfortable for some, it avoided heavy sanctions or coercion to ensure payment of the debt and prevented internal conflicts that could have damaged the group. Because the community had adopted a system to monitor what was owed to each member, all could see and understand the situation, follow the transactions, and be sure they were not getting cheated.

Salvatierra's leaders are still keeping the wage records but have stopped calling public meetings to present results and pay wages. The members of this group are less confrontational with their leaders, and since the process went smoothly at first, they have relaxed certain aspects that would generate more transparent public records. Salvatierra is a larger community, so it has been more difficult to gather everyone for meetings. As money from timber sales arrives, leaders have begun paying shares to individuals in need, rather than attempting general distribution. The individual payments are still documented and individuals receive receipts, but the group has become less consistent in posting records that list everyone's payments. The process is less transparent and there is less social control over the transactions because it has become hard to know what others have received. As long as there are no mistakes, they will not have problems. However, if errors occur, they will have no mechanisms that will allow easy resolution.

Lessons

These cases show successful monitoring that is locally useful and sustainable and contributes to good forest management. By adapting a process that already existed, we created a system that was easy to understand. The wages were an immediate, tangible benefit of the forest management plan. We simply needed to organize the information in a format that could be presented and understood by the rest of the community. The villagers knew what information was being recorded, how it would be used, and why it was significant to them.

It was easy to generate interest in monitoring because we began with an issue that was important to community members—their labor investment during the project startup and the resulting wages they were owed by the community management organization. The topic of wage payments held the attention of community members because the possibility of earning a wage was a major motivation for participating in management activities. As a result, they were willing to sit through lengthy group discussions and were genuinely interested in creating a system that would allow them to better understand and control the distribution of benefits.

The periodic payments offered a perfect opportunity to repeat the monitoring process, update the information, and evaluate it. Also, because the labor investments and payments formed a relatively short cycle, it was easier for locals to understand and evaluate. The public presentations and visual aids made the process transparent so that all could understand.

Our two cases demonstrate that community leaders are capable of consistently recording information for monitoring purposes. The information they gather continues to be used to ensure better administration of community resources in the future. Starting with records they already had certainly helped their confidence. Because members learned the importance of monitoring, they told us they were more concerned about recordkeeping in the field and worked to make sure all knew their totals.

The exercise increased local interest in monitoring, but it was driven by interventions coming from outside the community. The original recordkeeping system and the format for presenting and discussing the information were strongly influenced by the outside facilitators. However, the villagers have continued to use the system on their own. In the future, they will need to further adapt the system and develop similar systems for monitoring other aspects of their management plan. The skills they have gained through this experience will make such future expansions and improvements more likely.

Epilogue: Monitoring Wages for a New Project

Cururú did, indeed, adapt the wage-monitoring process to encompass additional aspects of their management plan. A clear example is the system they developed to invest the profits from their 2003 timber harvest in a community development project. The community members collectively decided to use funds (indirect benefits) to build three water wells. Previously, Cururú families had been drawing their water from the river.

To do this, they adapted the administrative monitoring system used for wage payments. In a single meeting, the community defined a budget for the project, approved a timetable, and designated a work team and team leader to build each well. Because of the time commitment required and the arduous work involved, the community decided to pay wages and provide meals to the workers. The team leader was responsible for tracking the days worked by each participant, as well as monitoring the food supplies provided to the cook. The team leader also organized receipts and submitted them to the management plan coordinator to account for spending. At the completion of the project, the coordinator presented a report to the entire community so that the workers' wages could be paid. These processes were initiated and carried out entirely by the community, without our assistance.

Within one month, a community that had never before built a water well had organized itself to provide three sources of safe, dependable water serving the entire village and designed the project to provide much-needed work for community members. The community members themselves acknowledged that this organizational capacity came directly from their experience with the management plan and their administrative monitoring.

CHAPTER 6

Tracking Broom Grass Resources for Equity in Zimbabwe

Richard Nyirenda and Witness Kozanayi

M AFUNGAUTSI FOREST, A PROTECTED FOREST IN Zimbabwe, was the stage for a recent process about resource sharing that involved the Forestry Commission and the communities living around the forest (see also Mutimukuru et al., this volume). Through this initiative, the Forestry Commission and the communities, via their elected resource management committees, have been collaborating in managing and deriving benefits from the forest. Fifteen committees have been set up around the forest, three of which have been the focus of CIFOR's work—the Batanai, Gababe, and Ndarire resource management committees (see Figure 6-1). With each committee, CIFOR researchers Richard Nyirenda, Tendayi Mutimukuru, Frank Matose, Jevas Sithutha, and Witness Kozanayi facilitated the formation of at least three resource user groups—for broom grass, thatch grass, and beekeeping.

One of the aims of CIFOR's research in Mafungautsi Forest is the development of collaborative monitoring arrangements. This chapter focuses on the monitoring work with the broom grass user group in Gababe Resource Management Committee. The example from this group, which set up its own collaborative monitoring system, shows how a clear, single-issue focus—in this case on equity—can help shape adaptive collaborative management (ACM) efforts, and in particular, monitoring efforts. Looking at a locally initiated example yields insights about how to facilitate collaborative monitoring in ACM processes that have external support.

Context

The broom grass user group at Gababe is not very different from the broom grass user groups at Batanai and Ndarire. Like the other two groups, it is composed of around 40 members, who move in and out of such user groups as and when their

Figure 6-1. *Mafungautsi Forest, Midlands Province, Zimbabwe*

interests and time permit. Also like the other two groups, the Gababe group comprises mainly women.

However, the Gababe group is the only one with a formal link to the resource management committee: the secretary of the committee is also a member of the broom grass group. It appears to be more coincidence than design, but it is significant that this committee has at least one forest user, and thus there is a direct communication channel between the committee and the broom grass group. This makes it easier for the users' concerns to be raised and taken up by the committee and has improved the level of collaboration between the group and the committee. In Gababe, the link between experiential information, decision making, and subsequent action that is essential in adaptive resource management is clearly made through direct personal contact. The other resource management committees consist of influential members of the community, none of whom are forest users.

CIFOR's action research in Mafungautsi Forest started in 2000 and included Gababe Resource Management Committee (see Table 6-1). Through an initial visioning and future scenario exercise facilitated by CIFOR researchers, the broom grass user group members analyzed the history of broom grass harvesting in their area.

They recalled that until about 1999, only a few broom grass collectors were active, cutting limited amounts of grass and leaving plenty uncut in the wetlands

Table 6-1. *History of Broom Grass Monitoring in Gababe, Zimbabwe*

Year	Activities
1994	• Pilot resource-sharing scheme initiated in Mafungautsi Forest.
	• Collaboration between the Forestry Commission and communities envisaged through resource management committees (RMCs).
1995–1996	• RMCs formed in communities living around forest.
	• Gababe RMC formed.
1998	• Permanent sample plot set up by Forestry Commission and RMCs to monitor sustainability of harvesting thatch, broom grass, and firewood.
	• Decline in grass productivity reported by communities and attributed to poor harvesting methods.
2000	• ACM research started by CIFOR in Mafungautsi Forest.
	• Gababe RMC selected for ACM research.
2001–2002	• Broom grass user group formed at Gababe.
	• Gababe broom grass group involved in visioning and scenario-building exercises.
2002–2003	• Grass-monitoring system implemented by Gababe RMC and broom grass user group.

and river valleys where the grass suitable for making brooms grows. The resource-sharing initiative, which started in 1994, stipulates that each broom grass collector must pay for a permit issued by the resource management committee before he or she can enter the forest and harvest the grass from the wetlands. So far, there had been no limit to the amount of grass each permit holder could cut, only for the duration one could cut, which was four days per permit.

In 2000 and 2001, the number of broom grass users increased, and thus the demand for grass also grew. The increase was largely due to economic hardships in the country in general. Some users began claiming very productive parts of the wetlands, which resulted in conflicts. The permit period was still four days, and the areas that many people were claiming were too big to be cut in four days. This problem was quite prevalent in a wetland called Wadze, one of the two wetlands involved in the resource-sharing initiative of Gababe Resource Management Committee.

Such issues were raised during a field visit organized by the Forestry Commission during a workshop held before the grass–cutting season and attended by all the resource management committees in Mafungautsi Forest. During the field visit, all the committees and Forestry Commission officials agreed that each committee should ensure that people had equal access to the broom grass resource. Determining the size of the wetlands and the amount of grass in those wetlands was proposed as a first step in trying to address the equity problem. Then the committee would be able to determine the unit area to be allocated per person per permit issued. Participants also observed that some committees seemed to include land with more grass than others, and thus a way had to be found to distribute the resource equitably both within each committee and across all the committees. After the workshop, Gababe Resource Management Committee resolved to initiate its own monitoring system aimed at addressing the equitable access issue.

The discussions helped the group identify several issues that could be improved, and for which it would need to work closely with the committee. One of these issues was ensuring that each member of the group had equal access to good-quality grass.

To ensure equal access, the group decided to monitor the size of the plot allocated to each permit holder and the amount of grass collected from each plot. Although they did not themselves refer to this as collaborative monitoring, it was clearly an example of a resource management initiative in which tracking changes could allow for adjustment of management practices as necessary.

A Plan to Ensure Equal Access

Assessing and Allocating the Resource Base

The Gababe Resource Management Committee started by determining the size of its *vlei* (wetlands). Committee members paced the river valley and estimated its size at approximately 50 hectares. They then divided the valley into equal plots of 2 hectares. A map of the valley was drawn to show the plots. The committee members then used wooden pegs to mark the plots on the ground so that everyone could see where the plots were located and where one could cut. Each permit holder was subsequently assigned a 2-hectare plot on which he or she could cut the broom grass, with each paid permit being valid for a two-day period.

Committee members patrolled to ensure that all those who were cutting had valid permits. They also made certain that grass harvesters were cutting only within their plots and were using the correct harvesting method—that is, with sickles and not the destructive digging up of the whole grass clump. The monitors also watched for spot or selective harvesting—a practice whereby harvesters take only the best grass from certain sections of the *vlei* and leave the poor-quality grasses. Selective harvesting had been prevalent in the past and was one of the main causes of conflicts among harvesters, since it caused a scramble among early harvesters for sites with good-quality grass. The late harvesters were usually the poor, who needed time to raise money for permit fees or those working their crops until the end of April.

The amount of grass harvested from each plot was recorded by a committee member during the harvesting. The annual productivity was calculated by summing the harvested grass quantities from all the plots. This was also undertaken by a committee member, who reported the findings to the broom grass group and the rest of the community. The committee used the calculations to report to the community on the amount of money they had raised through the grass–cutting permits. The broom grass group used the calculations to monitor the amount of grass they were harvesting each year, both total and per person. This provides trends on annual harvest and number of harvesters and helps them adjust plot sizes and allocations. For example, if the number of harvesters is increasing or the amount of grass harvested is decreasing, then plot sizes and numbers can be adjusted to keep ensuring equitable access to the resource.

Implementing the Monitoring

Initially, the new monitoring system was well received by the broom grass group. The grass cutters saw the new system as a fair way of ensuring access to their grass resource. However, implementation was not without problems. Gababe Resource Management Committee felt that dividing the wetlands into 2-hectare plots was an extra burden, given their other tasks. Thus only a few committee members volunteered for this task and for implementing the new monitoring system. The limited involvement made consensus difficult, and yet consensus was crucial for the new system to succeed. Consensus had to be sought on plot sizes and on the modalities of monitoring—who would do it, how, and when.

Some of the more influential members of the broom grass group tried to resist the introduction of the new system. These members saw that the new system would take away their advantage and tried to pressure some committee members to allow them to cut more broom grass than others. They were not successful.

Although the initial idea to create plots was designed mainly to ensure equitable access to the broom grass resource, both the committee and the broom grass group realized it also represented an opportunity to monitor the harvesting of the broom grass resource and thus assess whether levels were sustainable. However, there was only weak commitment from the user group itself to monitor other parameters that would reveal whether the quality of grass was improving or deteriorating over time. Nevertheless, the building blocks were now in place to monitor other parameters for other purposes, if they so wished.

Lessons

The Gababe broom grass group represents the first initiative by forest users in Mafungautsi Forest for implementing their own collaborative monitoring system. The group accomplished its aim at a time when the CIFOR researchers and Forestry Commission officers were grappling with how to introduce collaborative monitoring in the forest area. The Gababe initiative served as an example of how collaborative monitoring can be initiated locally in ways that address forest users' identified concerns. Several lessons can be drawn from this initiative.

Prior Collaboration

A history of collaboration in resource use appears to be a positive factor for initiating collaborative monitoring. The fact that the resource management committee and the broom grass user group had been collaborating before the monitoring system began made it easier for them to work together in designing and implementing the monitoring. Their prior collaboration had given them a set of common concerns, such as equitable distribution of the resource, and helped them reach consensus. In many other situations (including others in this volume), external stakeholders initiate collaborative monitoring with stakeholders who have no history of collaboration. Thus much time and energy are spent in encouraging cooperation among the stakeholders before collaborative monitoring can be undertaken. Collaborative monitoring arrangements developed in this manner certainly do not get off to an easy start. This lesson is reinforced by what happened in Gababe a year later: all the resource management committee members were replaced by people who were not forest users, and collaboration between the committee and the resource users ended.

Shared Monitoring for Equitable Access

For collaborative monitoring to succeed, it must deal with an issue that affects all the stakeholders. In Gababe, the issue of equity was affecting both the committee and the broom grass group. As a result, there was genuine shared interest in designing and implementing a collaborative monitoring arrangement that would help them deal with equitable access. When the plots were demarcated and allocated, the group members were happy that they could now all cut the same amount and quality of grass. This form of collaborative monitoring improved the relationship between the committee and the broom grass group, resulting in better collaboration. The earlier tension among grass harvesters arising from unfair access to the best grass areas was also defused, since selective harvesting ended with the introduction of the 2-hectare plots.

The committee and the group members began to engage more with each other as they tried to improve the system. They planned to monitor grass quality as well, by collecting samples from each of the plots. There are also additional opportunities for collaborative monitoring. For example, the committee could implement the same model of monitoring for other resources, such as thatch grass. Effective monitoring would reduce poaching of resources and institute an efficient system of recouping permit fees from resource collectors. There is also scope for the Forestry Commission to join in and assist with technical advice, depending on the outcomes of the monitoring.

Simplicity and Relevance

The collaborative monitoring initiative at Gababe is simple and relevant. It was initiated locally by two stakeholder groups. It did not require much data collection, since the plots were there on the ground for everyone to see and recordkeeping by the broom grass group was not necessary.

Most collaborative monitoring arrangements require considerable data collection and recordkeeping and usually involve a more complex set of stakeholders. By trying to satisfy the needs of many stakeholders at once, the collaborative monitoring efforts can quickly become overly complex and irrelevant.

A simple and relevant local monitoring arrangement, however, may form the basis for creating a larger-scale collaborative monitoring arrangement that deals with more issues and involves more stakeholders. In the Gababe case, others could get involved in measuring grass quality from the demarcated plots and recording the amount of grass harvested from each plot. Such information would help the Forestry Commission and other resource management committees track trends in the volume and quality of grass and thus adjust grass management strategies accordingly. Sharing of data, though it obviously would need to be negotiated, would not compromise the information needs of local stakeholders who are monitoring issues to ensure equitable access.

Epilogue: Ongoing Challenges

The Gababe Resource Management Committee secretary, an active broom grass collector, was voted out of office in a recent election. In fact, the elections had been long overdue—the constitution stipulates a biannual election calendar. The defeat of the secretary removed the crucial link between the broom grass collectors and the main committee. Not a single new committee member is an active broom grass collector.

In the subsequent grass-harvesting season, many conflicts ensued between the main committee and the broom grass group. The committee felt that the broom grass group was trying to usurp its power and dissolved the group. The broom grass collectors contended that committee members were unfairly allocating areas with high-quality grass. At one point, the conflict degenerated into a fistfight between a committee member and a grass collector. The case was reported to the Forestry Commission and the local chief. Though no fine was imposed, the disputants were strongly reprimanded.

Although the grass collected from each plot is still being monitored, what is missing is the crucial feedback loop into the community. No attempts have been made by the new committee to share the results of the monitoring exercise with the community. The new committee has now asked some of the old committee members to assist it, since the new members were not clear on how to run the organization and there was no proper handover from the old committee to the new.

We cannot stress enough the point that the effectiveness of monitoring depends on collaboration between stakeholders. The removal of the secretary clearly created a rupture between the committee and the grass collectors. Similarly, we feel it is significant that organizations formed without a clear role are bound to cause institutional problems with existing institutions. The feuding between the broom grass group and the main committee is primarily caused by the lack of clarity of the roles and linkages of the broom grass group to the larger organization.

PART IV

Dealing with Difference

CHAPTER 7

Monitoring with Strong Interests and Weak Incentives in Palawan, the Philippines

Herlina Hartanto

THE RELATIVELY LONG HISTORY OF COMMUNITY forestry in the Philippines began in the 1970s with government programs. Two decades of evolution in community forestry led to the establishment of an umbrella national program called community-based forest management. This program gained the status of a national strategy in 1996, through Presidential Executive Order No. 263, to ensure the sustainable development of forest lands in the Philippines.

The case study discussed in this chapter occurred in a community-based forest management area that covers three *barangays* (villages) and is located 67 kilometers from the provincial capital, Puerto Princesa City (Box 7-1, Figure 7-1). In 1996, the use rights of the 5,006-hectare management area were devolved by the Department of Environment and Natural Resources to a local organization called San Rafael–Tanabag–Concepcion Multi-Purpose Cooperative. The tenure agreement requires this local organization to protect, rehabilitate, and

---

Box 7-1. Community-Based Forest Management in Palawan

The management area consists of a strip of disturbed forest lands in need of rehabilitation and development. Prior to 1970, forest conditions in the area were good. A diversity of plant species was observed, with almaciga (*Agathis damarra*), ipil *(Instia bijuga)*, and narra *(Pterocarpus indicus)* as the dominant tree species. In 1970, pressure on the forests increased significantly with the operation of a logging concession and the in-flow of migrants who practiced slash-and-burn agriculture. As a result, the forest's condition started to decline. In 1986, a logging ban that included Palawan was imposed in the Philippines (Lorenzo 2001).

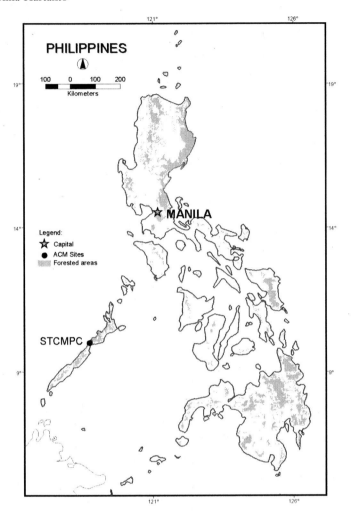

Figure 7-1. *Palawan ACM Site, the Philippines*

conserve the area. The cooperative must also help the government protect adjacent forest lands, prepare and implement management plans for the area, develop and enforce relevant policies, and follow laws, rules, and regulations pertinent to forest products use (DENR 1996).

The development of a local monitoring system in this case study was facilitated as part of CIFOR's adaptive collaborative management (ACM) initiative. The ACM research was carried out by two full-time facilitators from late 1999 to late 2002. This chapter is based on a two-year observation period, from February 2001 to April 2003, during which we facilitated discussions with several stakeholder groups in developing a local monitoring system for their community forest.

Context

At the onset of our ACM work, we encountered four significant challenges to community-based forest management in Palawan that affected the implementation of collaborative monitoring.

First, we encountered many government and civil society organizations engaging in community-based forest management, creating complex relationships and interactions. These groups held conflicting views on how natural resources should be managed, and coordination and communication among them were poor. Second, we noted low levels of citizen participation in community forestry implementation. The cooperative was led by a board of directors consisting of 12 members. Most of the cooperative's 433 members had joined the organization expecting to benefit through employment in rehabilitation initiatives or profit sharing or both. Their participation diminished quickly when easy direct benefits ceased. From the beginning, this cooperative had also been seen as a competitor by some of the *barangay* government officials and the non-timber forest products license holder operating in the area.

Third, the community and cooperative members had limited skills and experience in dealing with the complex procedural requirements of managing forest areas, collecting relevant information, and communicating them to other stakeholders. A fourth challenge was technical but had institutional implications: the rich biodiversity in the area. As one of world's biodiversity hotspots, Palawan has been subjected to many policies and regulations that aim to protect and conserve the remaining habitats. These had, and continue to have, significant implications for the extent to which natural resources can be used by communities.

Adaptive collaborative management was implemented in the midst of these challenges. As facilitators, we introduced ACM as an approach that could improve implementation of community-based forest management by enhancing the participation and engagement of relevant stakeholders and by making learning intentional and deliberate throughout implementation.

The need for monitoring was expressed by the cooperative. As a part of its responsibilities as a community-based forest manager, the cooperative was required by the local office of the Department of Environment and Natural Resources to undertake participatory monitoring and evaluation. Despite a commitment to do so, cooperative members had little experience and knowledge to develop and conduct such a system. They identified the lack of monitoring processes as a critical issue to address through an ACM approach, with our support. They stressed the importance of collaborative monitoring so that stakeholders would share costs and responsibilities and learn together. Besides our methodological support, they also sought monitoring support from government agencies, in terms of technical and analytical skills, data sharing, and finances.

The Monitoring Experience

The monitoring work in Palawan involved five major stakeholder groups: the cooperative, the Department of Environment and Natural Resources, the Palawan

Council for Sustainable Development, municipal and village governments, and local nongovernmental organizations.

We facilitated the development of a monitoring framework in three workshops and several discussion sessions outside the workshops. The first workshop was held in February 2001 and was attended by key personnel and members of the cooperative, two representatives from the village and city government, and two Department of Environment and Natural Resources representatives. The second workshop, in September 2001, brought together more stakeholders and community groups. Newcomers included representatives from the Women's Group, the Fishermen's Association, the Palawan Council for Sustainable Development, and three local NGOs. The third workshop, in January 2002, was attended by similar stakeholders as in the second workshop, except for the agency and the local NGOs, which attended only the preworkshop session.

The monitoring framework itself went through two iterations during these workshops before data collection started. The processes and steps of development of monitoring system are described in Figure 7-2 (see also Hartanto et al. 2002). They can be summarized into four major processes: developing the framework, searching for common ground, creating the monitoring arrangements, and data collection and learning.

Developing the Monitoring Framework

To facilitate the process for deciding what needed to be monitored, we introduced the criteria and indicators (C&I) framework for sustainable forest management. The choice of this framework was strongly influenced by CIFOR's prior work on criteria and indicators. The hierarchy of principles, criteria, indicators, and verifiers (Chapter 1) was developed to help organize information in a comprehensive and coherent manner (Prabhu et al. 1996, 1998b; Lammerts van Bueren and Blom 1997). In some cases, C&I had been useful in facilitating communication among stakeholders, including communities and other local stakeholders (De Oliveira 1999; Ritchie et al. 2000).

We asked participants to envision the ideal future conditions that they would like to achieve in the next 25 to 30 years. The components of their visions were subsequently broken down into smaller and more specific subcomponents. Through this process, a three-level C&I framework encompassing broad aspects of community-based forest management was produced.

At this stage, negotiation took place between cooperative members and representatives from other groups on which aspects of community-based forest management they considered important and should be included in the monitoring framework. In general, the cooperative was responsive to suggestions from other organizations on two conditions—first, that the inclusion of certain criteria and indicators be justified; and second, that those stakeholders provide support if the cooperative did not have sufficient capacity to monitor those criteria and indicators. By going through this process, the monitoring framework was refined several times. Input from community members who did not participate at this workshop

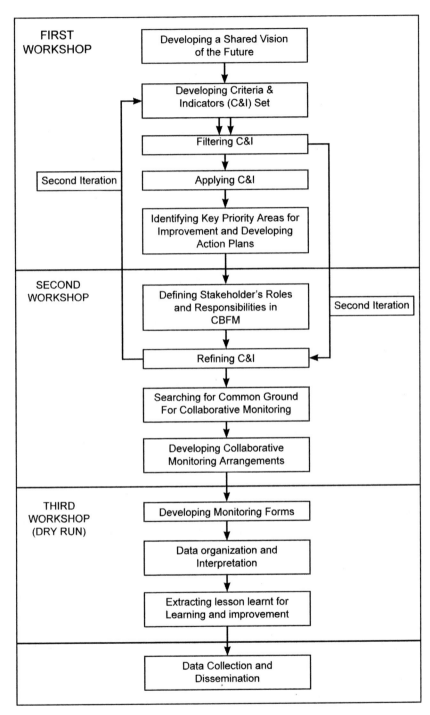

Figure 7-2. *Developing a Local Monitoring System in Palawan*

was sought by sharing the initial results through community newsletters and bulletin boards.

Finding Common Ground

We considered this part of the process critical, since a jointly developed framework does not necessarily mean shared responsibility for undertaking the work involved. Collaborative monitoring takes place only if there is a convergence of concerns and interests among the participating stakeholders. Thus, by relating the monitoring tasks directly to the fulfillment of organizational mandates, responsibilities, interests, concerns, and information needs, we expected the stakeholders to be motivated to participate actively.

It became clear in the second workshop how different the concerns and interests of the different stakeholders actually were. Government institutions, such as the Department of Environment and Natural Resources and the Palawan Council, were most interested in forest and coastal resource management and protection. The cooperative and communities were also interested in these aspects, but their main information needs focused on food supply, income, education, and health.

Despite the differences, there seemed to be sufficient shared interests and concerns among the stakeholders to form a basis for collaborative monitoring.

Creating Collaborative Arrangements

At this stage, participants discussed how to implement the monitoring system. The discussions focused on who would collect the information and how, whether some information existed already and where, how often the information should be collected, and how many data sets were needed to make the information meaningful. Later on, several members of the cooperative met to design the data collection forms. The C&I framework was revisited and appropriate forms combining several related parameters were developed. They field-tested the forms by interviewing personnel from organizations that had the information needed and refining the monitoring forms accordingly.

Unfortunately, the issue of how monitoring responsibilities would be shared between the cooperative and other stakeholders was not resolved. At the end of the second workshop, a memorandum of agreement was signed by representatives of the organizations that expressed a collective commitment to sharing results with others inside their organizations, especially the directors, to garner high-level support for the monitoring efforts. However, information sharing was only minimal. Hence the cooperative did not get the institutional commitment it had expected.

Collecting Data and Learning

Realizing that they had to initiate monitoring without support from the other organizations, the cooperative members started monitoring on their own. Understandably, they decided to monitor parameters of interest and utility for them—forest-based livelihood activities linked to community-managed forest areas, such as volumes of timber, rattan, resin, and honey collected, transported, and sold; and nonforest income-generating activities, such as goat production and handicrafts. They also monitored the incidence of illegal resource activities in the coastal and the community-based forest management areas, and they collected data on the number of students enrolled in the local school and the training sessions provided to their members.

Members of the cooperative started collecting data in February 2002 in line with their information needs. For example, most women were interested in monitoring livelihood parameters, such as handicraft and goat production, while the men were interested in monitoring the volume of forest resources extracted, such as lumber and resin. They then reflected on and analyzed the data collected.

In general, the process increased their awareness of how to improve their actions, improved the monitoring process itself, and guided their decision making on certain issues. For example, by recording illegal activities in their area, they realized that they did not know the appropriate mechanisms for reporting such activities or what information they should submit to facilitate quick responses from the responsible agencies. This awareness prompted them to learn more about reporting mechanisms for government agencies. In another example, the women monitored the time needed to complete handicraft items so that they could determine a reasonable price for the product that included not only the costs of the raw materials but also their labor. By recording which products were sold, the women gained insight into demand and were better able to plan production. They went on to monitor the price of similar products and the variety of designs available in the market. By comparing the profit they generated from selling rattan and raw materials vis-à-vis semifinished or finished products, officers of the cooperative realized that they should develop their skills to process rattan into furniture and other value-added products.

Unfortunately, what appeared to be a promising development in monitoring-triggered learning ended in April 2003. At this time, the Department of Environment and Natural Resources imposed a Philippine-wide ban on timber extraction, including the harvesting of salvaged wood from dead trees. The cooperative, whose income came mainly from wood extraction activities, was seriously affected by this ban. Several cooperative officers, including the chairman, had to seek economic opportunities elsewhere. Forest management activities were subsequently run by a skeleton crew of cooperative members, who were discouraged by the absence of their leader. Many members turned to seaweed planting at their coastal areas, a new income-generating activity. Harvesting of nontimber forest products fell tremendously, and the cooperative was unable to meet the targeted volume set for that year. Without these activities, monitoring stopped: there was nothing much to monitor.

Lessons

This case study highlights the tremendous complexities and fragility of setting up a multistakeholder monitoring system. Despite the convergence of interests and concerns and strong participation of stakeholders in the development stage, the cooperative members ended up carrying the data collection workload. They had attempted to gain the support and commitment from other stakeholders and, on several occasions, had also presented their monitoring system to high-level government officers. All these efforts were to no avail.

The absence of positive outcomes—in this case, no collective buy-in into the monitoring effort—may indicate that monitoring was not a priority for the other stakeholders. These stakeholders were accustomed to collecting data for monitoring by contracting with community members or assigning the work to low-level staff. This probably contributed to low involvement of these stakeholders in collaborative monitoring.

Another challenge that we faced in facilitating the development of monitoring in Palawan was the shift in stakeholders' concerns and interests. Early on, we had facilitated the development of a monitoring framework with the cooperative and other stakeholders from scratch. Our assumption was that this participatory process would increase ownership and commitment. Furthermore, no locally accepted C&I monitoring set was then available. However, in late 2001, the local Department of Environment and Natural Resources introduced a standard monitoring framework, called environmental performance monitoring, to community-based forest management managers in Palawan. Although the system was intended to be used on a voluntary basis, the local office planned to make it a compulsory monitoring tool as the basis for reporting and evaluating progress.

This new development led to conflicting information needs of the cooperative and agency: one system to guide the implementation of local resource management activities and another for upward accountability to authorities. These conflicting needs were not resolved, despite similarities between the two monitoring frameworks, and as a result, the Department of Environment and Natural Resources withdrew its support for the collaborative monitoring initiative. Although clarity about the commitment and contribution of each stakeholder is crucial, we now also see the importance of building stakeholders' capacities and willingness to renegotiate monitoring agreements and arrangements to accommodate evolving information needs, concerns, and interests.

As elsewhere in the Philippines, the skills and capacities of local organizations in Palawan need to be strengthened, especially in documentation and recordkeeping, and data organization and analysis. However, this was not the main stumbling block to our joint attempt at collaborative monitoring. The biggest obstacle faced by the cooperative is the conservation orientation of natural resources management in Palawan. This factor shapes how the Department of Environment and Natural Resources positions itself vis-à-vis local forest management efforts. Added to the inadequacy of livelihood options and development alternatives, that conservation objective limits the benefits that local organizations can obtain from community-based forest management. High fees, limited access to outside markets, excessive regulatory requirements, and the harvest permit ban have made forest management less attractive for local organizations in Palawan. This, in turn, affects their motivation for managing forests and monitoring in general, not to mention collaborative monitoring.

The many challenges in facilitating the development and implementation of monitoring at the local level are compounded by multiple stakeholders. Collaborative monitoring is likely to be sustainable if the participating stakeholders have clear incentives to take an active part. In Palawan, differing capacities, needs, mandates, priorities, and expectations manifested themselves as inter-organizational power differences, and attempts to set up multistakeholder collaborative monitoring did not take off as planned—the cooperative was the only active participant. Its autonomous monitoring efforts could have been sustained if community-based forest management had been perceived as sufficiently rewarding, but it was not.

Our process suggests that collaborative monitoring can work only under conditions of sufficient and differentiated incentives. Policymakers must reexamine the impacts of current policies on the socioeconomic conditions of local forest users so that more conducive policies and more supportive mechanisms can be established to genuinely improve people's livelihoods. Only with that incentive—the promise of a better life—will resource users invest in monitoring efforts.

The cooperative in Palawan is currently reorganizing itself. It will be interesting to observe how they will choose to pursue monitoring in the future.

Acknowledgments

The author would like to thank Ma. Cristina Lorenzo and Anita L. Frio for their contribution in facilitating monitoring efforts in this case study, and Azucena Gamutia for her assistance in collecting relevant information.

CHAPTER 8

Using Monitoring as Leverage for Equal Opportunity in Nepal

Cynthia McDougall, Chiranjeewee Khadka, and Sushma Dangol

*I*N MARCH 2002, "MS. B," a low-caste and impoverished member of the
Bamdibhir Community Forest User Group in Nepal, secured a position in a
promising local bamboo enterprise. Previously, this kind of opportunity would
likely have been inaccessible to her. However, a shift in the group's decision-making process toward adaptive and collaborative management (ACM) had created
an opportunity that Ms. B was able to seize. Part of the shift had entailed active
monitoring to hold the group accountable to its own equity-related goals.

In this chapter, we outline the monitoring-driven shift. Ms. B's ability to use
information as leverage to hold her group accountable to its decision illustrates
the potential for monitoring to support local governance that benefits the poor.
Her experience is a concrete example of the power of improved social process to
change access to forest benefits—and yet it is only one small part of the spectrum
of changes we witnessed in relation to social, human, and other forms of capital as
the ideas of ACM took hold in the forest user group.

Context

Ms. B's story took place within the context of a three-year action research process
in Nepal, from 1999 to 2002. The research, a formal collaboration of the Ministry
of Forest and Soil Conservation and CIFOR, was carried out in partnership with
two local research partners—the organizations New ERA and ForestAction—and
several independent consultants. As a research team, we sought insights into the
conditions, processes, and institutional arrangements for improving collaboration
and conscious social learning (Maarleveld and Dangbégnon, 1999; McDougall et
al. 2002a) in community forest management. Could forest managers, especially
communities, who based their management in collaboration and social learning

increase their effectiveness in sustaining and improving local people's livelihoods while maintaining or enhancing the resource base?

The research team combined traditional and participatory action research (PAR) in their approach to this challenge. By participatory action research, we refer to a process through which a group of people identify a problem, collect and analyze information, and act upon the problem to find solutions and promote social and political transformation (Selener 1997). Its foundation is a cycle of iterative learning: reflection, planning, action, observation, reflection. We distinguish PAR from research based on data collection and analysis conducted largely by researchers and oriented toward questions and variables identified by the researchers.

Our work at the local level focused on four case studies, including the Bamdibhir Forest User Group in Kaski District, discussed below (see Figure 8-1). The research teams in each site undertook background studies in 2000, followed by participatory action research in 2001 and 2002. We used traditional research in the background studies and final reassessments as a basis for assessing changes in each site over time (i.e., from 2000 to 2002) as well as for making cross-site comparisons. PAR was initiated to catalyze or enhance approaches focusing on adaptation and collaboration appropriate to the local situation, such as strengthening local institutions, addressing boundary negotiations, or increasing income-generating activities.

The nature of participatory action research implies that the research and action are designed around the social and biophysical context and oriented to address particular challenges. The section below highlights some of the contextual issues and challenges that shaped the research project framework and process in Nepal.

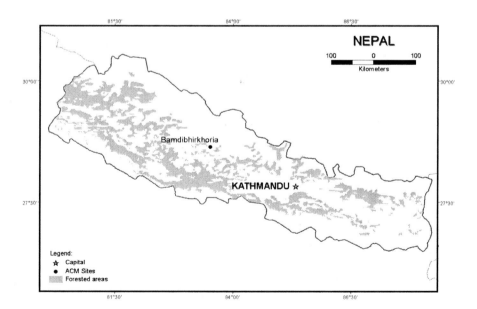

Figure 8-1. *Chanpakot Village, Kaski District, Nepal*

Nepalese Social Diversity

Even in the present time, caste shapes Nepalese society and influences social interaction, especially in rural areas (Bennett 1983). Power is distributed primarily along the lines of Hindu frameworks, in which caste and gender are driving forces. Consequently, participatory organization is difficult with regard to including the "excluded," such as lower castes, tribal peoples, women, and youth (Ojha et al. 2002).

Furthermore, as noted by Ojha et al. (2002) in the ACM project's comparative case studies, "although caste and ethnic differences do not necessarily determine matters such as the distribution of land ownership, income, consumption patterns and access to resources (Blaikie et al. 1980), almost all 'untouchables' are poor, and there is a high correlation between caste and wealth." Roles, responsibilities, and gender relations vary considerably with caste and ethnicity, making generalizations difficult; nevertheless, the overall pattern indicates that women bear greater labor burdens and have far less access to resources and decision-making processes.

The Need for Ongoing Innovation in Community Forestry[1]

The impressive record of the Nepalese government and civil society in establishing more than 12,700 formal community-based forest user groups in Nepal (Community Forestry Division 2003) is well known. However, although the Nepal Community Forestry Programme is reported to have improved forest cover in many cases, it has not yet clearly and consistently enhanced the livelihoods of all people dependent on forests in the community forest areas (Malla 2000, 2001; Springate-Baginski et al. 1999; Winrock International 2002; Kanel and Pokharel 2002). It is widely acknowledged that inequity in decision making within the forest user groups is common (Malla 2000, 2001; Springate-Baginski et al. 1999; Paudel and Ojha 2002; Winrock International 2002). As a result, relative to their needs, economically and socially marginalized peoples, such as women and low-caste groups, receive disproportionately small shares of any benefits that do emerge. Kanel and Pokharel (2002) even suggest that "in worst cases ... the implementation of [community forest] policy has inflicted added costs to the poor, such as reduced access to forest products and forced allocation of household resources for communal forest management with insecurity over the benefits."

In summary, our work identified several obstacles to generating benefits and achieving equity (see McDougall et al. 2002c). Specifically, we noted the following trends among forest user groups:

- Local elite tend to dominate internal decision making and benefit sharing while low-income and low-caste people and women tend to be marginalized.
- Decision-making processes and structures tend to reinforce existing patterns of marginalization.
- Planning processes tend to be linear and ad hoc in nature.
- Communication and information flow are weak.
- Management is often passive or narrowly focused on subsistence timber and fuelwood, with little emphasis on nontimber forest products.

- Legitimate users of community forests are sometimes excluded from the groups, and unresolved conflicts, such as over boundaries and benefits, are common.

Those trends occur in a very complex and dynamic context. Forests are an essential element of many rural livelihoods, and their use is intertwined with the use of other natural resources. Within forestry, there are overlaps, potential complementarities, and tensions between community forest and private forest use, as well as between forest user group members and nonmembers. Furthermore, the past 20 years has seen more stakeholders involved in all levels of community forestry, with shifting roles and policies regarding responsibilities and benefit sharing. The challenges of the situation are compounded by the limited human and financial resources for management. Kanel and Pokharel (2002) note in summary:

> This has meant a change in power and expectations of forest users and stakeholders, and that a new equation of social relationships among stakeholders has been started to establish [sic]. This has given rise to a need for strengthening collaboration among stakeholders so that they can negotiate, co-operate and devise appropriate institutional arrangements for resource conservation, management and use ... A need has emerged over the last few years for strategies that can add value to [community forest] processes and relationships so that equity and benefits can be enhanced.

The social and decision-making patterns we have described as challenges were evident in all the ACM research sites, including Bamdibhir Forest User Group.

Bamdibhir Forest User Group

Bamdibhir is located in Chanpakot VDC, Wards 3, 5, and 6, in Kaski District, in the Western Development Region of Nepal. At the time of the research, the forest user group had 134 households (722 people) as members. The members included a mix of ethnic and caste groups, with approximately 40 percent Brahmin households, 23 percent Magar, 21 percent Biswokarma, 9 percent Damai, 3 percent Chhetri, 1 percent Sarki, 1.5 percent Bhujel, and 1.5 percent Rai.

The level of families' dependence on community forest products in Bamdibhir is higher in the lower-income groups than in the wealthier groups. In a participatory wealth-ranking exercise that was part of the research process, the user group members developed categories. Of the total "poor" households in the forest user group, 30 percent are completely dependent on the community forest, whereas only 6.7 percent of the "wealthy" families fit this description (Khadka et al. 2003).

The 48.5-hectare forest area is a subtropical combination of natural and replanted forests. Its dominant tree species are katus *(Castanopsis indica)*, mahuwa *(Engelhardia spicata)*, and chilaune *(Schima wallichii)*, with uttis *(Alnus nepalensis)* in the planted area. This forest is a source of firewood, ground grass, fodder, and timber for building houses, shelters, and agricultural tools, as well as plants for domestic purposes, including medicinals.

Governance and Management at the Onset

The decision-making and governance processes in Bamdibhir Forest User Group were similar to those of many such groups in Nepal. The basic institutional structure consisted of an executive committee, including a chairperson, as the main decision-making body and an annual general assembly that was meant to be a forum for wider information sharing, discussion, and passing decisions.

Like many forest user groups, the Bamdibhir group operated in a fairly passive manner. The chairperson dominated the decision-making process, the committee and general assembly met irregularly, and members—especially the poor and low-caste people—had very little information about or input into the overall running of the group. Some conflict existed within the group between hamlets and ethnic groups and between the chairperson and other members. There were few activities and few successes, other than the effective protection and regeneration of some areas previously subject to landslides. See Table 8-1 for more information on institutions at the onset.

The Participatory Action Research Process

Our initial role as participatory action researchers was to offer facilitation of processes to enhance the members' capacity to adapt and collaborate within the group. For us, the starting point, and anchor, of the adaptive and collaborative approach was self-monitoring by the group. Thus we started the PAR by facilitating the development of an effective self-monitoring system. In all the four case studies, this process began with a local self-monitoring workshop.

In Bamdibhir, the workshop took place over three and a half days and involved 56 participants. The workshop focused on drawing out and integrating the perspectives of group members—visions of their ideal community and community forest, and assessments of the group's strengths and weaknesses. As well as joint visioning and developing and assessing indicators developed about and by the group, the workshop also included experiential exercises that highlighted the significance of learning and collaboration in forest user group management. The self-monitoring process during this workshop identified weaknesses that formed the basis for planning specific activities—income generation, forest protection, and reshaping of institutional aspects, such as distribution of forest products.

Participants (sometimes joined by researchers) then presented the workshop ideas, including the self-assessment and initial action plans, to Bamdibhir's four hamlets for further discussion and development with members who had not attended the workshop. The initial ideas for action were refined, and members formed "action groups" to lead them. This, in itself, already constituted a change in practice: in the past, the executive committee had led most if not all forest user group activities.

Several months after the implementation of the action plans, the self-monitoring process was revisited, thus beginning a pattern of iterative cycles of reflection and action (adjustments) in governance and management.

Table 8-1. *Comparing Institutional Structures and Processes for ACM Project in Bamdibhir Forest User Group*

	Pre-ACM project	Post-ACM project
Institutions and fora	• General assembly, supposed to be held every six months, had not been held for three years. Forest user group committee meetings often failed to meet quorum. *Tole* (hamlet) meetings for forestry issues were not held regularly.	• *Tole* meetings became major decision-making, planning, and conflict resolution fora. Each *tole* elected a subcommittee, whose monthly meetings fed into the executive committee's decision-making process.
Mechanisms for information sharing and input in decision making	• Members of forest user group had no mechanism to give committee input on community forestry management systems. General users knew little about committee decisions. • Monthly committee meetings and annual assembly provided theoretical mechanism for information flow. Information was also exchanged informally among members in tea-shops, gatherings, and workplaces.	• Main mechanism for input in decision making is the *tole* committee and *tole* representatives, who inform electorate of committee decisions and present *tole* decisions in committee meetings. Committee meetings are open. • Prior to decision making, *tole* users and committee members list agenda items and explain importance, discuss and prioritize issues, and set agenda. Most issues for discussion come from *tole* meetings. The goal is common understanding and consensus. *Tole* representatives now participate actively.
Mechanisms for conflict resolution	• Committee made some effort to manage conflicts over breaking of forest user group rules by members. Conflict between user group and chairperson was not addressed.	• Mechanisms to resolve conflict exist. Conflicts within committee and between it and *tole* members can be discussed in *tole* meetings and resolved.
Mechanisms for distribution of resources	• Decisions about distribution of resources were made by committee or chairperson with little input from, or consideration of, marginalized groups.	• Committee bases decisions about distribution on agreed criteria developed through its discussion, through discussions in *toles*, and through heterogeneity analysis system.
Access to training and sharing of experiences and learning	• Information about workshops and study tours was not shared, denying group members the opportunity to participate. The first person to hear about the chance would grab it (or push forward a close relative). Chairperson dominated decisions about access to training, which only committee members attended. No formal system for sharing experiences and learning after participation existed.	• Committee formally selects participants for training, workshops, and study tours after seeking information and decisions from *tole* members. • Committee has criteria for membership that include gender, caste/ethnicity/class, and *tole*.

Note: McDougall et al. 2002b.

Innovation in Governance and Management

As described above, based on the action plans that emerged from the initial self-monitoring workshop and discussions, the forest user group members and committee gradually implemented shifts and innovations in the institutional arrangements and planning processes. Table 8-1 outlines five related areas of innovation: institutions and fora; mechanisms for information sharing and input in decision making; mechanisms for conflict resolution; mechanisms for distribution of resources; and access to training and learning. These shifts were enabled by group members' improved facilitation and participatory process skills, supported by both formal training and informal support from the research team.

In this section, we focus on the changes in decision making and planning because we see these as major building blocks for increasing marginalized members' access to opportunity. Initially, although the forest user group had decision-making bodies (the executive committee and general assembly), the process of decision making and planning was far from systematic. Annual plans and activities were generally made and carried out at the discretion of the chairperson of the executive committee. In the rarely held general assemblies, the committee (mainly the chairperson) would present the proposed plans, but only the more vocal members participated in discussions, which tended to focus on current problems.

During the PAR process, Bamdibhir Forest User Group planning became more systematic and more inclusive through three intertwined developments. First, self-monitoring became the basis for planning, including for annual work plans, and the renewed constitution and operational plan. Thus decisions and plans were clearly linked to future visions and based on critical reflections of past progress.

Second, heterogeneity analysis—looking at differences in well-being, occupation, institutional role, forest dependence, and demographic factors—helped the forest user group "track and assess who was participating, contributing, and benefiting . . . and trends in participation, contributions, and benefits for marginalized versus non-marginalized members proved critical . . ." Thus quantitative data enabled both a quantitative and a qualitative analysis of equity. The executive committee, with the help of members and researchers, developed a system for this analysis based on current information about group members, including *tole* (hamlet), ethnicity and caste, education levels, and wealth ranking. Heterogeneity analysis proved critical for Ms. B.

Third, based on critical reflection on weaknesses in governance during the self-monitoring process, the forest user group implemented strategies to make the decision-making process more transparent and inclusive. Specifically, major decisions started to be made through the participatory self-monitoring process, *tole* discussions, and committee discussions with *tole* representatives. For example, the *tole* committees and the executive committee jointly began to develop the agendas for committee discussions and general assemblies, and the committee and *tole* representatives together started finalizing agreed-upon action plans for implementation based on *tole*-level decisions. The general assembly became more of a forum for final approval of plans rather than the main opportunity for members' input.

The Bamboo Craft Enterprise

In Bamdibhir, the initial self-monitoring process undertaken by the forest user group illuminated weaknesses in several areas, especially income-generating activities and equitable distribution of information and benefits. One result was the development of a bamboo craft enterprise. The agreed-upon goal of the enterprise was the "economic upliftment" of marginalized group members, and especially women. With the help of a nongovernmental organization called Bamboo Secret, training was organized for 16 members.

After the training course, the Monitoring and Evaluation Team of the Bamboo Craft Training, which consisted of committee members and Bamboo Secret representatives, selected five trainees to become paid staff in the enterprise. The Monitoring and Evaluation Team stated that its recommendation of the five had been based on the goals of the enterprise and an agreed criterion—that each had earned a passing grade in the course. The recommended list of paid craftworkers was accepted and was about to be approved when one participant, Ms. B, spoke out.

As a lower-caste, poor woman, Ms. B would not have had much, if any, leverage to influence committee decision making in the past. In this case, however, by explicitly referring the group to the monitoring-based decisions and tools, she was able to make her point effectively. Ms. B pointed out that the recommendations did not meet the previously defined and agreed-upon goal of the bamboo craft enterprise—to provide opportunity to marginalized members, especially women. She referred to the forest user group's heterogeneity analysis, the self-monitoring tool that had been used to identify which families fell into the target group. Based on the records, she fit the criteria of "marginalized" better than others who had been selected. They fell into the "lower middle" and "poor" categories, whereas she was "very poor." Furthermore, she had received a B grade in the training course.

Following Ms. B's statements, the committee, Bamboo Secret representatives, and meeting participants reconsidered the decision and appointed Ms. B as a paid employee in the bamboo craft enterprise.

Self-Monitoring: The Key to Accessing Opportunities

Although Ms. B's newfound ability and courage to speak up in a public forum may be related to many factors, including participatory action research in general, we view her success in obtaining a position as very likely linked to self-monitoring and the larger ACM approach (of which self-monitoring is an essential part). There are several reasons for this.

First, we note that the bamboo enterprise itself was a direct outcome (action plan) of the self-monitoring process, which identified income generation as a weak area of the forest user group. Second, the self-monitoring process identified marginalized members as being in greatest need of income-generating activities, and the group confirmed its intention to give these members priority—hence the criterion that employment in the bamboo enterprise benefit marginalized people.

Third, the reflections in the self-monitoring process flagged the need for increased transparency, accountability, and inclusion in decision making, including regarding equity issues. The resulting heterogeneity analysis tool made explicit who was in the economically poorest category of the forest user group and triggered more explicit communication of rationale for decisions by the committee—for example, the Bamboo Enterprise Committee publicly shared the goals of the enterprise and the criteria used to select employees.

We view the two last points as especially significant in Ms. B's case. As part of the self-monitoring, the forest user group had explicitly recorded the goal of income opportunity for marginalized users and, through the related heterogeneity analysis, identified exactly who fit that category. Rather than having to argue about fairness in the abstract, Ms B. could present data generated from a planning process based on self-monitoring—and prove that a decision was inconsistent with the established criteria.

Lessons

This story resonates with the larger trends we witnessed across the four case studies in the research project in Nepal. In governance and management, we saw a shift from ad hoc and undocumented goals and decisions to explicit goals and decisions developed through iterative visioning and a self-monitoring process. Emerging from that, we also saw a shift from committee-dominated decisions to decision making and planning that involved the active input of members.

Each forest user group increasingly operated as a broad-based learning forum by developing a shared vision and indicators, undertaking self-assessment and monitoring, and basing action plans on the assessment. As this evolved, we observed not only more input from marginalized users but also the identification of more governance weaknesses, especially regarding equity of access to decision making and benefits. This recognition that governance was a weak area tended to enhance the transparency as well as the accountability of governance, helping leaders stay on the "agenda" of improvement under the increasingly watchful eye of the members. Thus the self-monitoring process created incentives and pressure on the committee to enhance its accountability.

The same can be said for equity issues. Forest user group members made equity more explicit as a goal by including it in the joint vision and in the indicators. The reference to inequity in specific action plans appears to offer marginalized group members an opportunity for leverage. In the Kaski research sites, this was heightened by the use of heterogeneity analysis as a monitoring tool.

The changes in decision making and planning were catalyzed by the PAR process. Although the forest user group members were the main actors, the research team played the important role of change agents or catalysts in several ways—by highlighting the significance of equity, collaboration, and shared learning in community forestry management; facilitating the initial monitoring and participatory decision-making processes; and training and backstopping members and community forest support agents in facilitation and self-monitoring processes.

In sum, these innovations, based on adaptive collaborative management with self-monitoring at their center, appear to allow forest user groups to become more inclusive and begin to adjust power differences. In our experience, this created opportunities for the marginalized users to be heard, in some cases for the first time. Social capital appeared to increase with the increased satisfaction of most members regarding access to decision making and its transparency. The changes also translated to a shift in the distribution of human capital in the group, in terms of access to information and skills through training. It is too early to know for sure, but the shift toward increased access for marginalized forest users to information, influence, and opportunities may help lay the foundation for increases in well-being for the poorest of the forest-dependent poor.

Adaptive collaborative management and related self-monitoring processes are not silver bullets for development and natural resource management. Many hurdles must be jumped en route to successful implementation of such processes, and even then, policy, resource, and economic barriers remain. Nevertheless, examples like that of Ms. B offer some hope that these processes may contribute in some significant way toward more equitable and effective natural resource management and development.

Acknowledgments

This chapter is based on CIFOR's research project "Planning for the Sustainability of Forests through Adaptive Collaborative Management," funded by the Asian Development Bank (RETA 5812). Cynthia McDougall was the project leader for the Nepal research and Chiranjeewee Khadka and Sushma Dangol were research team members. The project lessons that contribute to this chapter emerged from the hard work and commitment of the research team. Besides the authors, these were Mani Banjade, Hemant Ojha, Krishna Paudel, Raj Kumar Pandey, Bharat Pokharel, Shibesh Regmi, Him L. Shrestha, Netra Tumbahangphe, Laya Upreti, Hima D. Uprety, Kalpana Sharma, and Narayan Sitaula. We would also like to thank Dr. Keshav Kanel, Mr. K.B. Shrestha, and Dr. Bharat Pokharel for their contributions to the project. Last, but definitely not least, we would like to acknowledge the commitment and contributions of the forest user group members in Manakamana, Andheri Bhajana, Deurali-Bagedanda, and Bamdibhir, the staff of the District Forest Office and Federation of Community Forestry Users in Kaski and Sankhuwasahba, and the multiple other individuals and organizations at the local, district, and national levels that engaged in and enriched the project.

Note

1. This section is drawn from McDougall et al. (2002c) and McDougall (2003).

CHAPTER 9

Monitoring to Ease Forest Management Conflicts in Cameroon

Phil René Oyono, Mariteuw Chimère Diaw, Samuel Efoua, William Mala, and Samuel Assembe

*I*N 1994, CAMEROON LAUNCHED A COMPREHENSIVE restructuring of its forestry policy that had hitherto been dominated by government agencies. Driven by devolution and decentralization, many management rights and responsibilities were transferred to other actors—local communities and communes (local governments)—allowing them to create community forests and council forests and to have direct access to forestry revenue (Carret 2000; Milol and Pierre 2000; Fomété 2001; Bigombé Logo 2003). A new zoning plan was also designed to define regimes for forest access. The paradigm of adaptive collaborative management (ACM), based on the interactions, collaboration, and mutual learning of stakeholders (Lee 1999; Borrini-Feyerabend et al. 2000), made it seem relevant and innovative for forest management systems under these policy conditions. Thus ACM research was planned for Cameroon with its new institutional developments and strong demand from stakeholders for more participation, collaboration, reciprocal adaptation, and horizontality (Ebene and Oyono 2000; Diaw and Oyono 2001).

The ACM research team undertook collaborative monitoring efforts in five sites: Lomié, Dimako, Ottotomo, Campo, and Akok (see Oyono et al. 2003a) (Figure 9-1). These sites are large management areas—a cluster of five community forests, a communal forest, a state forest reserve, a national park, and a landscape mosaic that hosts many communities and forest stakeholders. This chapter focuses on Dimako, Lomié, and Akok and describes different processes of learning, each shaped by the resource issues that local stakeholders face.

Located in the east of Cameroon, Dimako has a council forest—a forest classified for or planted by a local government and belonging to the private estate of that government. The classification lays down the boundaries and objectives of the forest's management. Lomié, also in the east, covers five community forests. A

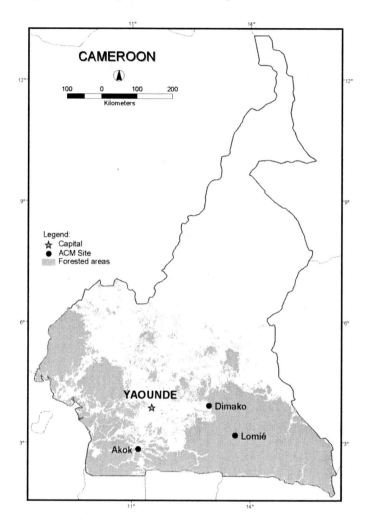

Figure 9-1. *Dimako, Lomié, and Akok, Cameroon*

community forest belongs to the national estate and is subject to a management agreement signed between the village community and the forestry administration. Akok is the site of a landscape mosaic—agricultural and forest land—in the humid forest zone of Cameroon.

The goal in each of the three sites was to create a multistakeholder process that could generate a sustained and effective mechanism for managing forest regime changes and related institutional transformations. ACM team members worked as facilitators and data and information managers with a range of stakeholders that are specified in each of the three cases below. The ACM work lasted from early 2001 to the first quarter of 2003.

Dimako: Competing Claims

Dimako's management plan was designed in 2000 by the French-funded Forêts et Terroirs (Forest and Lands) Project. Demarcation of the council forest had ignored customary landownership, and land-use conflict emerged. Two villages toward the north of the council forest were claiming seven square kilometers of agricultural land that they said had been snatched from them by the council forest and over which they asserted historical rights. The council responded that forestry legislation gave forest ownership to the state. Thus the issue in Dimako was rooted in intergenerational access to the forest and different claims to rights and power over the forest.

First, the ACM team members started a participatory action research (PAR) process involving all major stakeholders—the people of the two villages, council authorities, and administrative authorities. The first PAR meetings enabled local stakeholders, the researchers, and project experts to better understand the problem and identify its causes and consequences. Then, the ACM team organized several meetings to facilitate joint problem solving (Oyono et al. 2003a; Assembe and Oyono 2004).

The two main protagonists—the villages and the Dimako rural council—both resorted to legal arguments based on customary and modern law. Nevertheless, they agreed to a dialogue facilitated by the ACM team. After many rounds of discussion, council authorities and the forest management committee agreed to reinstate local rights over half the area taken from the two villages. In addition, the people of another neighboring village, Ngolambélé, agreed to give part of their secondary fallow land, which was not included in the council forest, to people of one of the affected villages, Nguinda. These decisions were the main outcome of the PAR process and formed the basis of subsequent collaborative monitoring.

On-site stakeholders drafted an agreement covering all the stages of the retrocession process. An ad hoc committee comprising local community representatives, the municipal council, the Technical Operational Unit of the Ministry of Environment and Forests, and the council forest management committee was created to follow up the process. To seal the plan, an official statement was signed by the stakeholders and witnessed by administrative authorities (Assembe and Oyono 2004).

After the agreement was reached, the ad hoc committee defined how change would be implemented and determined that monitoring should focus on the retrocession process and the allocation of the additional fallow land to the two villages.

Over a period of 10 months, the ad hoc committee and ACM team monitored the process of change through various meetings, using the agreement and social and agroecological maps. Information related to the two main action areas. Thus meetings focused on how and when the council was returning the reallocated agricultural land to the two villages, on how the neighboring village was ceding its old fallow land, and on how the two villages were sharing the returned land.

Problems arose when the two villages had to share the old fallow lands ceded by Ngolambélé. One village wanted to take the lion's share, arguing that it was of the same lineage as Ngolambélé. When the ACM facilitators and the ad hoc committee suggested instead that the village population levels should determine the division, that idea was accepted. In November 2002, the retrocession process was successfully concluded, in part because of the collaborative monitoring efforts. Since then, the role of the committee has been to establish and implement local mechanisms aimed at sustaining change, and the monitoring focus and responsibilities have been adjusted accordingly.

Lomié: Local Governance at Stake

In the Lomié region, newly created and established community forests had triggered a "forestry fever" centered on access to revenue. This generated various conflicts:

- between NGOs and local communities, on what the latter called "excessive supervision"(Klein et al. 2001; Efoua 2002; Etoungou 2003);
- within local communities, on how their community forest should be used, some favoring small-scale logging for sustainability reasons, an option defended by NGOs, and others favoring the more profitable medium-scale logging through logging companies; and
- within local communities, on the mismanagement of revenue by local management committees, which created interest in monitoring organizational arrangements and ecological sustainability.

Stakeholders included local communities, the local office of the Ministry of Forests and Environment, the Federation for the Development of Community Forests, logging companies, and two NGOs, one local and one Dutch.[1]

During a strategic meeting organized in January 2002 with the federation and the local NGO, the participants drew up a program of action focused on exchanging experiences among the five community forests of the Lomié site. The idea was to strengthen village management committees and share information about negotiating timber sales, using income generated from any sales, and handling any related conflicts.

During a forum bringing together all stakeholders and the ACM team, the federation and the local NGO received a mandate to monitor this process of capacity building within the committees with the methodological support of the ACM team (Oyono et al. 2003a). The local communities were involved in the monitoring through community forest management committees, which belonged to the federation. Information about forest management and any related income and conflicts was collected from each village. Monthly on-site meetings were held to share experiences, discuss community forest management, and explore best practices leading to both community forest sustainability and human well-being (Box 9-1). These village-level meetings brought together on average 30 participants.

In parallel, the ACM team facilitated the collaborative development and simplification of indicators of sustainable community forests management and of human well-being (Table 9-1). Two local meetings organized by ACM and the Dutch NGO in February 2001 brought together representatives of the five villages managing community forests, representatives of the federation, the local NGO, administrative authorities, and a timber company working with the five villages. During these meetings, participants were asked to identify signs showing that well-being

---

Box 9-1. Information Gathered through Monitoring and Related Analysis and Action

In Kongo, the community contracted with a town-based operator to log its forest. The management committee undertook the negotiations. Timber is cut twice a month and pays a reasonable return. The revenue is mainly invested by heads of family in housing improvements. No internal conflicts have emerged as yet.

In Koungoulou, the community forest was exploited by two logging operators simultaneously, in a confused situation marked by intensive and uncontrolled logging. The price of timber was badly negotiated by management committee members who fought only for their own interests. No positive socioeconomic outcome ensued, since the money was spent on food and drink.

The ACM team, the local NGO, and the Federation for the Development of Community Forests invited representatives of Koungoulou—including management committee members—and those of Kongo to meet and exchange experiences and strategies. These meetings led people in Koungoulou to select better practices by working with only one operator and establishing accountability rules for committee members.

Table 9-1. *Indicators for Monitoring Community Forests in Lomié*

Principles	Indicators
Community forest well-being is effective	Many trees
	Animal species conserved
	Rationally exploited (for a long time)
	Abundance of nontimber products
	Good production of pharmacopoeia
	Abundance of fish in rivers
	Abundance of fauna
	Presence of moabi (*Baillonnella toxisperma*) in the forest
Human well-being is effective	Fair and adequate use of revenue generated by commercial exploitation
	Internal conflicts related to the management of revenue are resolved
	Sanctions applied for personal use of revenue
	Relevant community organization
	Secure access to resources
	Many sources of income
	Support from NGOs
	Many young people
	Social harmony
	Many children
	No rural exodus
	Health centers
	Functional schools
	Enough food
	Strong houses
	Electricity
	Many wells

was present in their community forests, and those showing that human well-being was present in these villages.

During a forum organized in March 2001, all stakeholder groups agreed that these simplified indicators would serve as a framework for conflict resolution, problem solving, and monitoring of community forest sustainability. Their "box" of indicators is being used by communities to monitor local forest management and benefits accruing from the forest resource.

To use the box, they transform all the indicators themselves into objectives of the sustainability of their community forests and of their own well-being. These simplified indicators come with instructions—written by both the Regional Forestry Service and the community—as part of the simple management plan for each community forest.

Useful monitoring results were obtained in two villages. Management committees in these villages, supported by an ACM facilitator and an NGO technician, reported on a monthly basis to the federation. Meetings were then organized with local communities to analyze information, identify parameters of positive change, and draw lessons to be shared with other villages. These meetings mainly considered the question of how many indicators seemed to be suggesting progress. The conclusions were that much time would be needed.

These collaborative monitoring efforts continued until January 2003, when ACM-Cameroon activities, including monitoring, ceased. In anticipation of this phasing out, the local NGO had been closely involved so that it could assume ACM team roles. Meetings were planned with local communities to continually adjust the simplified indicators of ecological and human well-being.

Akok: Shifting Cultivation Versus Innovation

Collaborative monitoring in Akok involved local communities, the International Institute for Tropical Agriculture, a French agricultural research organization,[2] and farmers' organizations.

Forests are shrinking in southern Cameroon because of agricultural expansion (Oyono et al. 2003b). Searching for a long-term peaceful coexistence between local agricultural practices and forest sustainability is imperative. When the ACM team launched its activities in the Akok site, members encountered a complete lack of interaction among stakeholders, notably between local communities and the research institutes (Oyono et al. 2003a). The former perceived technological change as just another extravagance of conventional research, and the latter considered shifting cultivation a historically backward behavior by people who had little concern for the landscape impacts of their annual slash-and-burn practices.

In 2001 the ACM team, with the agreement of the International Institute for Tropical Agriculture, the French agricultural research institute, and local farmers, started to promote collaboration by generating a more common understanding of the problems between the two groups. The ongoing discussion that this generated has so far enabled the research institutes to proceed with more collaboration from farmers, the intended end users of their research findings.

The next step involved developing local indicators of sustainable forest and agricultural areas (Mala 2002; Oyono et al. 2003a). Two on-site meetings were organized, in November 2001 and in March 2002, for the identification of these local indicators (Table 9-2). The first meeting brought together 23 participants, including 15 farmers, plus ACM facilitators and representatives of the agricultural research institutes. The second meeting drew 17 participants, including 12 farmers. Farmers were asked to describe signs that their agricultural activities were not destroying the dense forest and signs that they were active participants in the decision-making process, including reflection and discussion. The signs, or indicators, were chosen after long debates facilitated by the ACM team. During the first meeting, a list of local indicators was established. The second meeting revisited these signs before producing a final list (Table 9-2). The objective of this set of indicators is to help farmers monitor the impact of slash-and-burn agriculture on the forest ecosystem, and thus to reduce negative effects.

Local communities and the ACM team are monitoring the indicators in Table 9-2 using focus group discussions and observation. Data are collected by two local enumerators, representatives of the two agricultural research institutes, and the ACM team. This group then also analyzes the information. Thus far, management adjustments have focused on the first principle, through sharing information on the cropland-forest interface and more involvement of women in discussions about slash-and-burn agriculture. Information collected thus far is being used by local communities, the agricultural research institutes, and the ACM team to assess how agricultural space and forest space are interacting (Robiglio et al. 2003).

The last stage, which is in progress, has involved sharing the "Co-View" tool with farmers, agricultural researchers, and NGOs. Co-View is a computer-based planning tool through which stakeholders agree on a vision of a problem and find a common strategy of what to do and how to do it. The set of previously identified indicators serves as input for Co-View in creating a common vision of the future of the forest (Mala et al. 2003).

Table 9-2. *Indicators for Collaborative Monitoring in Akok*

Principles	Indicators
Local communities' participation in decision making is effective	Relevant information sharing
	Improvement of women's status
	Relevant interactions between research institutes and farmers
Landscape integrity is preserved	Existence of dense forest for present and future generations
	Many trees
	Existence of multipurpose trees
	Many home gardens
	Availability of agricultural land
	Existence of fertile soil
	Many animals
	Many fish

Lessons

Each of the three research sites tells a story about collaborative monitoring shaped by the issues at hand. In the Dimako site, the issue—retrocession of agricultural land—was important locally, and the associated information collection and analysis had great local significance. Even small steps in forest management, such as a change in ownership of land, had to be monitored carefully. Written agreements were very important in the monitoring process in Dimako, where all the parties signed a document defining the points to be followed during the retrocession of land, and therefore were obliged to respect their commitment. The document provided a common basis for the subsequent monitoring work.

In Lomié, the management of the community forests and resolution of the resulting conflicts had local and national importance. Since 2001, community forests have been created throughout the country. As the first to be managed, Lomié offered crucial lessons in collaborative monitoring, and outside observers followed this community's experience and data collection. The additional scrutiny helped keep the monitoring process on track, identify lessons for local communities elsewhere, and document the issue of community forest management in Cameroon. Akok's issues had both local and national relevance. We learned from this site that it is difficult to implement a real collaborative monitoring process in the field of technology change. In many cases, monitoring priorities are defined by agricultural research institutes that maintain exclusive control of data collection and analysis.

To be successful, collaborative monitoring efforts must be rooted in each stakeholder's concerns and interests. By definition, stakeholders rarely share the same initial interests. For example, the on-site timber company did not have the same interest in problem solving or collaborative monitoring as the local communities. Even within local communities, management committee members may not always have the same interests as other community members (Chapters 5 and 8).

Collaborative monitoring must therefore be approached creatively, as a search for equilibrium between stakeholders and their sometimes competing interests. In this balance, collaborative monitoring must have three qualities: retrospective (concerned about the past), prospective and investigative (oriented toward the future), and anticipative (capable of dealing with uncertainty) (Diaw and Kusumanto 2003). The three stories provide evidence of this. The retrospective aspect is highlighted by the situation in Dimako, where stakeholders were looking back to the past, to the former village boundaries. The investigative and anticipative aspects—apparent in all three sites—involved the cyclical phases based on observation of change and on actions oriented to sustain change in the long term.

To summarize the lessons learned from our initial efforts in Cameroon, we, the facilitators of collaborative monitoring, offer a metaphor used in various research sites to convey its essence:

> Let us think of a church built with very simple, local material. The priest and his flock want it to remain in a good condition during the rainy season as well as the dry season. If the roof is damaged by wind or rain, they will

combine their efforts and solve the problem, as they will for damaged walls or doors. But after damage to the church is repaired, the priest and his flock will not sit back with crossed arms. They will continue—retrospectively, prospectively, and anticipatively—to collaboratively unify their efforts in order to sustain the church.

It is the same with collaborative monitoring.

Epilogue: New Initiatives to Monitor

Logging operations in the Dimako Council Forest were launched in February 2004, on 2,400 hectares. A portion of the revenue that will be generated is intended for community development in the two villages. The ad hoc committee set up during the PAR process has met twice with municipal authorities to define the type of small-scale projects to be undertaken. In early September 2004, both villages received about US$3,600, following the decision to spend a portion of the revenue on school fees for the best students.

Meanwhile, following suggestions by ACM and the International Institute for Tropical Agriculture, farmers in Akok have launched a broad-based farmers' movement in the whole zone, including neighboring villages. This initiative generated a farmers' federation, the Union of Common Initiative Groups of the Akok Area. Its mandate includes, inter alia, the sharing and dissemination of indicators of collaborative monitoring in the area.

Notes

1. SNV = Stichting Nederlandse Vrijwilligers, a Dutch development organization; CIAD = Centre International d'Appui et Développement, a local development NGO.

2. CIRAD = Centre International de Recherche Agricole pour le Développement, the French development research organization.

PART V

Adapting Monitoring Processes

CHAPTER 10

Improving Forest Beekeeping through Monitoring in Chimaliro, Malawi

Judith Kamoto

L OCAL FOREST USERS REPRESENT SIGNIFICANT COLLECTIVE human resources, or social capital, in terms of knowledge of their environment, organizing capabilities, understanding of local institutions, and abilities to develop forest management systems that deal effectively with complexity and surprise. Around the globe, however, their social capital has consistently been underappreciated and underutilized. Government forest departments seem stuck in "command-and-control" forest management regimes. Forest policy makers have been slow to respond to changing circumstances, and their institutional mechanisms often interfere with the goal of achieving sustainable forests and communities. As a result, both forest quality and human well-being have suffered.

In Malawi, villagers around the Chimaliro Forest Reserve undertook to monitor sections of the reserve in return for certain usufruct rights. The agreement to cooperate in managing the forest has presented challenges to both the national forest agency and the local communities, but through their collaboration, ongoing improvements in resource management are becoming apparent. The effort hints at the potential of local forest users' social capital.

Context

In the Chimaliro Forest Reserve of Malawi (Figure 10-1), local communities had been denied access to the forest reserve since it was established in 1921. In 1992, the communities entered into a comanagement partnership with the country's Forest Department, which stipulated that they be allowed to collect nontimber forest products and hang beehives for honey production in exchange for regular patrols of the forest reserve, early-season burning to reduce fuel loads, and firebreak maintenance. In 1996, the Forestry Research Institute of Malawi was given the task of developing a management plan for the Chimaliro Forest Reserve. Three blocks were demarcated for the comanagement partnership, each having a historical connection with the group of villages under which it would be managed. The areas of blocks I, II, and III were 18, 118, and 74 hectares, respectively.

Development of this first plan was preceded by technical analyses by scientists from the Forestry Research Institute of Malawi. Although they consulted local communities, they produced models and guiding principles for management with little active stakeholder participation. This plan was, therefore, to a large extent an internal affair of forestry and ecological experts. Subsequent implementation was characterized by noticeably weak involvement of stakeholders. The plan provided little inspiration for communities, since they had not participated fully in its development.

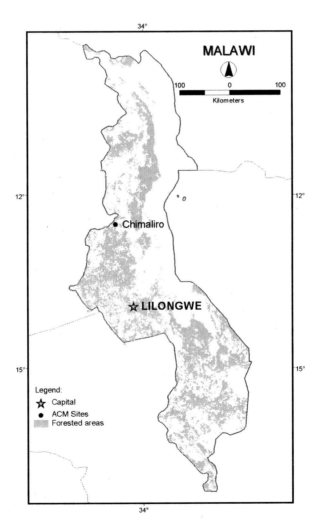

Figure 10-1. *Chimaliro, Kasungu District, Zambia*

In 2001, a comanagement agreement was signed between the Forest Department and the villagers around the Chimaliro Forest Reserve, who were represented by their local institutions, known as block natural resource management committees. Each block had its own committee. A benefit-sharing mechanism from proceeds of sales of confiscated illegal goods was set up, with 70 percent for the Forest Department and 30 percent for the block committees.

Nevertheless, illegal harvesting of honey, timber, and poles, to which communities were still officially denied access, has continued. Harvesting of timber and poles for home construction remains a contentious policy issue, since legalization requires policy changes. The agreement has had other implementation problems as well: the comanagement activities have benefited the committee members more than the communities they are supposed to represent.

To overcome these problems, the communities, led by their local institutions with facilitation support, tried an adaptive collaborative management (ACM) process. My role, as facilitator, was to guide the communities through an analytical process that included visioning, scenario building, and vision implementation and used the principles and methods of participatory action research, participatory rural appraisal, and collaborative monitoring. The core idea was to adapt management in a conscious and continuous manner by facilitating widespread use of self-improving and equitable forest resource management systems that build on local capacity, with "vertical" and "horizontal" stakeholder interactions.

Four major players had roles:

- Process facilitators documented the ACM process, collected data on the committees' activities and contributed to the discussions during reflection meetings.
- Forest guards led patrols, organized reflection meetings with committee members, and communicated with the process facilitator, committees, and other district forestry office staff on meeting dates and venues, and also had other government-assigned duties.
- Community members participated in block activities.
- Committee members participated in block activities.

In applying ACM at Chimaliro, we also developed collaborative monitoring to overcome some of the problems highlighted above. This chapter describes our progress with the monitoring work beginning in April 2002 and focuses on beekeeping, which is an important forest resource. Although this experience shows how specific aspects of monitoring, like the use of indicators to stabilize the use of beehives in communal areas, can be useful, our primary insights relate to the need for continual assessment of how monitoring can be improved and thus remain effective in generating forest management improvements.

Initial Monitoring and Honey Thefts

Before the comanagement agreement, patrolling the forest reserve was the duty of government-paid forest guards, and harvesting of honey was illegal. The agreement made individual block committee members partly responsible for patrols of their blocks and allowed honey collection as one of the benefits. Each block committee, with the Chimaliro forest guards, would set aside a day per fortnight for patrols. Normally, the chairperson, treasurer, a committee member, and the forestry staff would patrol to see whether the beehives were colonized and estimate the time remaining until harvesting.

Other monitoring activities included checking on thefts of honey and illegal tree felling. When the hives were full, the treasurer and other committee members and the forestry staff would harvest. The honey harvested never reached the market, however; the individuals involved in harvesting mostly shared it. At times, the treasurer, who was a custodian of the harvesting equipment, would secretly harvest the honey and blame the theft on the local community and outsiders (community members not involved in comanagement). Such acts caused considerable conflict among the committee members, but there were no monitoring indicators to assist the committee and the forestry staff in documenting the problem.

The block committee members reacted to the thefts of honey by removing the beehives from the blocks and hanging them in individual homestead wood-lots. However, the Forestry Institute of Malawi subsequently demanded that the beehives be returned to the forest blocks, since their removal defeated the very objectives of forest comanagement. No mechanisms were established to tackle theft problems, however.

Adjusting the Monitoring Plan

When the ACM process started, each block committee and a few community members from each block participated in a visioning exercise organized by the researcher. This visioning was based on scenario building, an exercise that aims to stimulate creativity and can be useful when complexity and uncertainty are high (Wollenberg et al. 2000). We used "force field analysis" (Box 10-1) to tease out the forest management problems within the community.

During the scenario exercise, bush fires, illegal tree felling, low honey productivity, thefts of honey, weak collaboration between block committees and the forestry

---

Box 10-1. Force Field Analysis

Participants reflect about their current situation and the kinds of problems they face, visualize the problems, and draw them on paper. They are then asked to draw a picture of their desired future. Participants compare both pictures and discuss the forces that encourage or discourage changing from the present condition to the desired one. They use this understanding of the positive forces (e.g., resources available) and negative forces (e.g., constraints) affecting their goals to strategize about the best actions to take to accomplish their goals. These actions need to be consistent with the forces, so that actions counteract negative forces and reinforce the positive ones.

Source: Adapted from Narayan and Srinivasan 1994.

staff, and limited involvement of the wider community in block forest management issues were highlighted as forest management problems. Central to the discussion were the low productivity and widespread thefts from beehives. Production of honey is affected by illegal felling—hives are often relocated so that particular trees can be cut—and by fires that destroy nectar-producing flowering trees and vegetation. The final product of the scenario exercise was a plan of action: indicators were needed to help stabilize use of beehives in common areas.

Based on ACM ideas and following the plan of action, monitoring indicators were developed by the wider community and block committee members, with three process facilitators, chosen with assistance from the Forestry Research Institute of Malawi. Two of these process facilitators had previously worked with the institute on other initiatives but were not on a government payroll. They were trained by the author to undertake the facilitation role and document the collaborative monitoring process.

The block committee members, a few community members, and the forestry staff agreed that monitoring should still be conducted by patrols but should be guided by the indicators. Using indicators could help track changes from the start of the intervention, in this case arresting those taking honey from the beehives as well as improving hive productivity. Indicators could help assess the causes of any observed changes. Box 10-2 contains an example of beekeeping monitoring indicators identified by actors of each block.

Ongoing Improvements through Critical Assessment

Interblock Cooperation

The ACM approach to monitoring via patrols at Chimaliro was carried out at the block level. Each block committee and the other actors undertook the patrols as independent entities, not as partners. However, at the reflection meetings, even though they were not called regularly (Box 10-2), all the blocks met, and the block committee members would share their patrol experiences.

At one meeting, participants realized that none of the committees were having much success in arresting the honey thieves. Members would point fingers at each other whenever illegal tree felling, honey thefts, or bush fires in their blocks were noted by a patrol. A coordinating committee was therefore formed to oversee issues of all three blocks. Membership was drawn from the three block committees.

As time passed, the monitoring data showed continued low production and thefts of honey and illegal cutting of trees in Chimaliro. The coordinating committee recognized that block-level patrols were inadequate. The patrols had been able to identify the culprits from within the committees and the wider community, but other factors prevented the apprehension and arrest of suspects: the patrols had previously relied on forest guards and local chiefs for conflict resolution and had not become accustomed to their new roles.

The coordinating committee proved to be the key to the solution because it represented all three blocks and could therefore approach and apprehend a culprit

---

Box 10-2. Collaborative Monitoring Agreement for Beekeeping

Local planning objective (identified by the actors below): to obtain honey and wax that can generate income for household food security and reverse the trend of privatizing common pool resources (the beehives).

Actors involved:
- block committee members;
- the Chimaliro forestry staff (forest guards); and
- several community members.

Indicators (identified by the actors above):
- availability of beehives and wax at all times;
- frequency of patrolling the beehives;
- number of hives harvested, quantity of honey and wax produced;
- time for harvesting;
- retail and wholesale price for the honey harvested;
- market availability and income realized from marketing;
- handling of honey and income: the treasurer of the block committee stores and sells the honey and banks the cash; and
- use of income generated for community development.

Means of data collection:
- patrols;
- transect walks; and
- research partners (the process facilitators), secretaries of each block in the field recording and documenting the information.

Means of sharing and reflecting:
- ad hoc meetings attended by the actors above.

---

from any other block. After reflecting on block-level patrols, the coordinating committee decided to embark on joint patrols in the reserve, scheduled weekly, then biweekly. During these patrols, all participants would check the beehives and look for felled trees or fresh stumps and any signs of fires—all potentially harmful for honey production.

Seeing the Merits of Resource Assessments

Further reflections on patrols made the coordinating committee and other actors realize that scheduled patrols seemed to have little effect in curbing illegal activities, including honey thefts. They believed that the culprits studied their movements and were able to continue stealing. The coordinating committee therefore decided to switch from regularly scheduled patrols to equally frequent but unscheduled patrols. This change did reduce honey thefts and illegal tree felling.

To improve honey production, the ACM researcher decided to introduce the concept of participatory forest resource assessment to support the collaborative monitoring of resources. Such a resource assessment involves all the community's stakeholder groups in undertaking a joint assessment of the forest resource and its condition, use, and potential. The stakeholders analyze the data as a basis for discussing the condition of the forest resources and its future management and then develop a forest management plan together. The existing forest management plan for the Chimaliro Forest Reserve, developed by the Forestry Research Institute of Malawi, provided little inspiration for the communities, since they had not participated fully in its development. A management plan developed through a participatory assessment would, we hoped, give the community a greater sense of ownership.

The participatory resource assessment exercise was conducted during five days in Block II. What we learned about developing forest management plans at the block level could later be used in working with the other blocks.

The assessment exercise made the block committee members, several community members, the process facilitators, and the Chimaliro forestry staff aware that certain forest management decisions had been ill advised. For example, beehives had been located in an area with forage for livestock but few mature trees for bees. They were also located far from water sources. Beehive location had contributed significantly to the poor colonization of the hives and, hence, to low honey production. Participants agreed to move the beehives in their blocks closer to a water source to increase chances of bee colonization and honey production.

The forest resource assessment also helped the people of Block II realize what resources to monitor where. Participants identified areas for thatch grass collection and areas to monitor illegal cutting of trees based on the availability of trees for different uses. They found that the trees on the upper slopes of the Chimaliro Mountain were suitable for construction poles and firewood for beer brewing and brick making, but were at risk from high demand. They also realized that grazing animals harmed regenerating stems of the most valuable species on the lower slopes. They agreed to regulate grazing and avoid bushfires in these parts.

Improved Management

Subsequent monitoring has provided interesting results. For example, the beekeeping monitoring has shown considerable improvement in communal harvests of honey. For the first time on record, Blocks I, II, and III harvested 18, 17 and 20 liters of honey, respectively, in two harvests from the reserve and the village woodlots. The honey was sold at Kasungu district headquarters for 150 Malawi kwacha per liter (about US$2 in 2002).

This achievement has triggered considerable growth in the number of privately owned beehives—from one to five in Block II and from three to seven in Block III. Although the block committees do not feed the information from monitoring back to the wider community whom they represent, the information is shared through other existing channels, such as general social interactions between individuals and at group gatherings like funerals, community-based training, etc.

The participatory forest resource assessment experience also taught the block committees about the conditions required to produce honey, leading them to adapt their practices. The clear benefits accruing from this nontimber forest product stimulated a wider community interest and willingness to learn about the resource base.

Collaborative monitoring has improved forest management in several ways that were not possible based on the initial block patrol system. A conscious ACM-driven process led to more effective patrols and a coordinating committee. A stakeholder-owned forest management plan has been developed. And those involved in patrolling can identify new illegal practices, such as garden encroachments on the borders of the reserve.

Nevertheless, participants in the ACM process continue to identify weaknesses—for example, poor participation of the wider community in joint patrols. Several problems with sustaining the monitoring-related learning process remain. For instance, the data on honey production are still collected by facilitators because individual beekeepers have difficulty recording the information. To maintain transparency and accountability, the monitoring indicators need to be updated, now that other agents (district forestry officers) sell the honey and the cash is banked by the chairperson, treasurer, and secretary.

Lessons

Monitoring can, indeed, trigger learning when it brings understanding about how the resource is changing, why these changes might be occurring, and what they signify. In Chimaliro, the information that was collected and shared fed the analysis that led to ongoing improvement to resource management. Developing indicators and identifying specific issues to monitor guided the communities, especially the block committees, in discovering the value of monitoring collaboratively. The process of creating the collaborative monitoring mechanisms enabled block committees to set goals for sustainable beekeeping, honey production, and forest management, besides helping them assimilate critical information about the effects of local forest management activities. It also contributed to the assessment and evaluation of progress toward the goals they had set for their forest reserve.

Local ownership of the collaborative monitoring process is critical. When individual blocks conducted their patrols independently, little or no success was recorded. These patrols had been set up as an obligation under the comanagement agreement. If communities are to take ownership of information systems to collectively make and implement decisions, then they must be participants in developing the monitoring process.

As summarized by Prabhu (2003), the goal of the monitoring should be jointly agreed on and clear to everyone involved. Critical questions to consider at the onset include these:

- Who made the decision to start developing a set of indicators? Is there community ownership of this fundamental decision? That is, does everyone agree that it is a good idea, and is everyone committed to it?

- Are the indicators going to be generated and used only by the community itself, or in collaboration with other partners (e.g., in a comanagement relationship)? In either case, is other assistance, such as facilitation, needed?
- Who will take responsibility for which parts of the process? Who will do the monitoring?
- Who gets to use the information and how?

Without a clear understanding of the overall purpose and final use of the information, participation will be compromised and confusion may occur, leading to later disappointment.

CHAPTER 11

Initiating a Dynamic Process for Monitoring in Mafungautsi State Forest, Zimbabwe

Tendayi Mutimukuru, Witness Kozanayi, and Richard Nyirenda

MAFUNGAUTSI FOREST OFFERS A VARIETY OF resources, including pastures for grazing animals, thatch grass, broom grass, medicinal plants, honey, mushrooms, firewood, construction timber, game meat, edible mopane worms *(Imbresia belina)*, indigenous fruits, and herbs. This 82,000-hectare forest is located in Gokwe Communal area, Midlands province, Zimbabwe (Figure 11-1). The majority of stakeholders around Mafungautsi belong to two main ethnic groups, Ndebele and Shona, with the rest belonging to four other ethnic groups—Shangwe, Chewa (immigrants from neighboring Malawi), Tonga, and Kalanga. Except for the Shangwe, the rest of the ethnic groups are immigrants to the area. It is in this area that CIFOR's adaptive collaborative management research is being carried out (also see Chapter 6).

Context

Mafungautsi Forest became a state forest in 1954 because it is a source of four rivers that flow to the Kariba Dam, a major source of hydroelectricity in Zimbabwe. Like all the other state forests in the country, its management was, and still is, vested in a government forestry agency, the Forestry Commission. Many conflicts arose from the change of tenure around the forestland. These conflicts grew in intensity and compelled the Forestry Commission to embark on a resource-sharing experiment in Mafungautsi.

The resource-sharing project was initiated in 1994 and went beyond the provisions of the existing Forestry Act by allowing communities to harvest some minor products legally, such as broom and thatch grass, mushrooms, and firewood, in return for which they helped manage the forest. To enhance community participation in the management and use of the forest resources, the Forestry Commission initiated the formation of resource management committees in all the communities

Figure 11-1. *Mafungautsi Forest, Midlands Province, Zimbabwe*

surrounding the forest. A total of 15 such committees were set up in the communities surrounding the forest.

However, even though the Forestry Commission tried to enhance community participation, progress was very limited, in part because of the lack of a joint monitoring system to enhance local learning. Monitoring continued to be solely the responsibility of the commission's Forest Protection Unit. This monitoring involved patrolling and policing and arresting people who transgressed the forest laws; it focused on controlling use of the forest rather than learning.

CIFOR's research initiative in Mafungautsi began with an effort to initiate and facilitate collaborative monitoring systems that would enhance social learning processes as part of the experimental project. The five steps we followed are outlined below, each offering important lessons.

Initiating Collaborative Monitoring

Step 1. The Survey: Understanding the Context

First, we undertook a preliminary survey to identify existing monitoring arrangements and ascertain the general flow of information and knowledge among stakeholders. This would help us determine how best to intervene and facilitate collaborative monitoring as an improvement to the existing monitoring.

The survey showed that at household and community levels, monitoring was fragmented. Different stakeholders were monitoring different things, according to their interest, without necessarily sharing their information with others. At the household level, people were largely concerned with the level of output from their fields; for those involved in harvesting forest resources, the income generated from the sale of the resources was important. At the level of the resource management committee, the main focus was on income and production of forest resources every season. Individual user groups were also monitoring different resources from the forest: for instance, broom grass collectors were monitoring the amount of grass collected by all group members to assess the group's performance at the end of every grass-harvesting season (Chapter 6).

There was, however, no systematic documentation of information gathered or any deliberate attempt to reflect critically and act on the information obtained from the monitoring processes. Some members of the resource management committees were not certain why they were monitoring what they were monitoring; the activity was simply undertaken as an unquestioned official task of the committee as mandated by the Forestry Commission under the auspices of the resource-sharing program. The user groups were not sharing or analyzing information.

Moreover, each resource management committee consisted of only seven members who were responsible for all the work: issuing permits to resource harvesters, ecological monitoring, and promoting beekeeping and tree-planting projects. Most of these duties were carried out ineffectively: there were too few committee members to share the work burden. The formation of subcommittees (see below) changed this situation.

Lessons. Facilitated collaborative monitoring can build on existing monitoring systems. Such systems probably exist in every community, in many forms (cf. Estrella et al. 2000), but normally go unrecognized since they are regarded as part of local people's daily activities. This was the case in Mafungautsi, where local people did not label their monitoring activities as such. After these existing monitoring systems have been identified, subsequent work can build on them and help all stakeholders make them deliberate and systematic.

Step 2. The Framework: Negotiating Understandings

Having identified the existing monitoring mechanisms, we produced a framework of how collaborative monitoring could be initiated and facilitated in Mafungautsi. We discussed this with the Forestry Commission officer, who provided additional comments and helped us understand his perception of monitoring. According to him, collaborative monitoring meant tracking the ecological condition of the forests through patrolling, for which the Forest Protection Unit was responsible. He envisaged an increase in the number of people patrolling in the forest.

There was no element of collaboration in that view of collaborative monitoring: the officer mainly saw the advantage of free labor for extra patrolling of the forests.

Our own perception of monitoring was that it should (1) involve monitoring both ecological conditions and human well-being, and (2) be truly collaborative and encourage stakeholders to share their monitoring results, reflect on them, and learn together to improve their management strategies and their livelihood.

After discussing the matter, we agreed on a definition of collaborative monitoring.

Lessons. The Mafungautsi case showed that negotiations to define collaborative monitoring are central during the initial phases of such a monitoring process. The Forestry Commission official and the facilitators had different views of collaborative monitoring, hence it was crucial for the two parties to discuss and agree on a definition.

Before initiating any collaborative monitoring processes, we had spent considerable time thinking about the best entry point. The context studies gave us a clear understanding of the local power dynamics in Mafungautsi, and we decided to work with existing organizations rather than create new ones and flood the organizational landscape. We identified the Forestry Commission as the best entry point, since we wanted its support and it was a major stakeholder in the resource-sharing project. Coincidentally, as we were initiating the monitoring processes, the Forestry Commission officer had just elected monitoring subcommittees that were supposed to assist the resource management committees in doing their work. Each subcommittee consisted of seven members.

In addition to thoroughly investigating the existing monitoring arrangements and understandings, it is important to unravel the local power dynamics and configurations to avoid misalliances. Several dangers lurk that could jeopardize collaborative monitoring: overburdening groups with new responsibilities, concentrating power in a few hands, or selecting groups that lack local credibility.

Step 3. Focus Groups: Discussing Stakeholders' Perceptions

After agreeing on the way forward with the Forestry Commission officer, we held discussions with many key informants—resource management committee members, traditional leaders, longstanding chairpersons of other resource management committees, and members of the newly formed monitoring subcommittees and resource user groups. Participants discussed different understandings of monitoring, which we used as the basis to negotiate a shared, operational definition of monitoring.

Discussions with the monitoring subcommittee revealed that the members saw monitoring as policing and arresting people. When asked about their role in a collaborative monitoring system, they said that it would be exactly like that of the Forest Protection Unit—that is, patrolling the forest and arresting people who broke the laws. They welcomed the idea of introducing a collaborative aspect to local monitoring because they anticipated the power and authority that they could assume in the system. This was not the case with resource user groups, who were initially against the idea of making monitoring more collaborative, since they associated it with an increase in punishments and fines.

At the end of the discussions with the various groups and further discussion with all stakeholder groups, alternative perspectives on monitoring emerged that were more positive. We explained that monitoring meant not just policing but could also include activities that were more directly beneficial to local people, such as keeping track of ecological and social changes, learning from them, and adapting in order to improve. During the discussion, participants also offered ideas on the kind of collaborative monitoring system that could be set up and the role of the monitoring subcommittee. In all the discussions, stakeholders indicated that the monitoring subcommittee would steer and drive the collaborative monitoring process.

Lessons. The monitoring history of stakeholders affects the way they view and understand collaborative monitoring. In Mafungautsi, resource user groups initially rejected the idea of introducing a collaborative monitoring system, based on their understanding of the term. Negotiations were necessary to convince local stakeholders that monitoring involved more than just policing and could be useful and beneficial to them.

Collaborative monitoring can be introduced not as a new concept but as something that will add value to what stakeholders are already doing. The facilitators used examples (Box 11-1) that helped the stakeholders understand the importance of monitoring to keep track of where one is going and to stay on track to achieve one's desired goal. Unless stakeholders are convinced that they stand to benefit from collaborative monitoring, they clearly have no incentive to participate sincerely in such a process. Time must be invested to achieve this understanding.

Step 4. Community Meetings: Agreeing on First Steps

Findings from the discussions with key informants were presented by the researchers to the community, whose views on the proposed collaborative monitoring system were solicited. All the local stakeholders (including the traditional leaders, the resource user groups, and women's groups) were present. The meetings began with various stakeholder groups' definitions of collaborative monitoring. This was followed by a description of the system as defined by the key informants, in which a monitoring subcommittee would play a central role.

The participants at the meetings were generally satisfied with the new definition and decided to adopt the proposed monitoring system. In this system, each stakeholder group—resource user groups, the resource management committee, traditional leaders, the Forestry Commission, monitoring subcommittee,

---

Box 11-1. The Potential Relevance of Monitoring

If you want to go to Gokwe Town from Mafungautsi Forest, there are several things that you look for while traveling in the bus to be certain that you are actually headed for Gokwe. Such landmarks include signposts, buildings, and vegetation. If you watch for these landmarks but do not see them, then you should quickly get off the bus and find the one that is going to Gokwe. If you do not check for familiar features, then you may end up lost in an unfamiliar place.

researchers—was supposed to monitor what it was interested in. For example, the resource user groups would monitor thatch grass, broom grass, honey, and poles.

The monitoring subcommittee was given responsibility to spearhead the monitoring process and help organize and coordinate platforms for the stakeholders to share their findings. It was supposed to continually interact with the resource user groups and assist them (Box 11-2). The terms of reference highlighted the linkages between the monitoring subcommittees and other stakeholders and the ways in which stakeholders could share information, reflect, and learn together. The roles of the various stakeholders were also clarified in the same discussions.

Lessons. Stakeholders need opportunities to discuss and deal with their fears. In Mafungautsi, we made fast progress once people's initial fears about collaborative monitoring—that it meant more policing for even trivial violations and would intensify acrimony between the community and the Forestry Commission—were dispelled. It then became possible to set up collaborative monitoring processes.

It was easy to introduce the idea of collaborative monitoring once the stakeholders had built good relations with one other, had developed a level of trust, and were willing to work together. At the onset of the research project, this would have been impossible: stakeholders were hostile toward one another and did not want to collaborate in any way.

One team needs to spearhead the collaborative monitoring process, especially where multiple stakeholders are involved in monitoring different things at different levels. Stakeholders in Mafungautsi wanted someone to take a lead role in enabling the different stakeholders to share results of their monitoring and suggested that the monitoring subcommittee could take this role, thereby building on existing institutions instead of creating new ones.

Stakeholders' roles must be clarified to avoid conflicts and duplication. In Mafungautsi, stakeholders were concerned that if the roles of the monitoring subcommittee and the Forest Protection Unit were not clear, conflicts would arise. Terms of reference were developed to define the roles of the monitoring subcommittee and its relationship with the Forest Protection Unit and the other stakeholders. This was also an attempt to avoid duplication of effort.

Step 5. Stakeholders' Workshop: Finalizing Terms of Reference

At the end of the design process, a district workshop was held to finalize the terms of reference of the monitoring subcommittee and to update traditional leaders and representatives of other resource management committees outside the research area. At that workshop, a draft was presented by the facilitators and members from the resource management committees that had participated in writing the terms of reference, and this was finalized. In the presentation, we stressed that this was just one example of a collaborative monitoring system, intended as a guide for developing similar systems—it was not a prescription. Stakeholders were encouraged to adopt or adapt collaborative monitoring systems that suited their contexts. After the workshop, some resource management committees set up monitoring systems that suited their local conditions, using

Box 11-2. The Terms of Reference for the Monitoring Subcommittee

The monitoring subcommittee should work in the forest, in the community, and also outside the community, and their responsibilities are as follows:

In the forest, to observe and look out for:

- forest fires, and notifying the resource management committee if there is a fire;
- people who cut live trees (and poles), and reporting them to the committee;
- poaching of wild animals, broom grass, domestic animals;
- snares;
- theft of livestock (some cattle rustlers were said to steal cattle and pen them in the forest while finding markets; the committee would therefore patrol the forest looking for such thieves);
- protecting the forest from fire by checking if the fire lines are maintained (if the committee discovered fire lines that were not maintained, it could notify the committee, which would then notify the village heads, who would then take the appropriate action);
- making sure that neighboring resource management committees do not go beyond their boundary when harvesting resources; and
- investigating the committee's work to make sure they are not stealing resources or fraudulently giving out permits to people from outside their area before the set time for harvesting resources.

In the community:

- investigating if people involved in any project are benefiting from it;
- checking if projects are doing what they planned to do;
- linking up all the other subcommittees with the resource management committee;
- making sure that the money raised through payment of fines when someone is caught by the monitoring subcommittee goes into the resource management committee's account;
- checking and observing if the other subcommittees, notably project committees, are working well toward the attainment of the group's vision (there was a need to monitor how projects were being run so that advice could be given if there were problems and also lessons learned could be shared with the larger community); and
- showing those who do not know areas where various resources can be harvested and advising the committee and the other subcommittees on areas where resources can be found.

Outside the community:

- The subcommittee should have a relationship with similar subcommittees in other resource management committees and visit them to find out about how they are managing their resources and check the quantities they have for specified resources. If there is a shortage of a certain resource in their area, then the monitoring subcommittee, together with the resource management committees, can ask the Forestry Commission to expand their harvest to those areas with plenty of resources.
- Also, if they see someone from other resource management committees breaking the rules, they need to hand over such culprits to their respective committees.

(continued)

Box 11-2. *(continued)*

Reporting structure, frequency of feedback meetings and records:

- Who needs what data, and how often should the information be reported and in what format? It was recommended that literate people be involved who could document the results of the monitoring exercise and share them with others. Monitoring without written reports would be inadequate, since people would want information on what was happening in their area. Who needs the data collected by the subcommittee should be spelled out, as should how often he or she wants information presented. A recording format for the information should be agreed on.

- What is the reporting structure when people who break rules are caught by the subcommittee? Stakeholders suggested that the monitoring subcommittee fall under the resource management committee; the subcommittee should have no direct link with the Forestry Commission. All issues should be sent to the main committee, which will in turn hand over cases they cannot deal with to the traditional leadership authority. If traditional leaders are unable to deal with a case, they can hand it over to the commission. However, stakeholders said that the commission should feel free to talk to the subcommittee at any time. But if the resource management committee is being implicated in a report, the monitoring subcommittee can report directly to the traditional leaders. According to the stakeholders, the subcommittee should also arrest offenders from outside and refer the person to his or her committee for prosecution.

---

the terms of reference as guidelines. Other committees, however, adopted and implemented the example system.

Lessons. To reduce the complexity of a large collaborative monitoring effort, start with a small but representative group. Developing a guideline for collaborative monitoring is cumbersome because it involves multiple stakeholders with divergent interests. A small group can develop a collaborative monitoring system that will later be presented to a wider group for comments. Scaling up the collaborative monitoring process can later be done by giving the pilot group the opportunity to share the concept with other stakeholders—but only, of course, if they themselves are convinced that a collaborative monitoring system is important and beneficial.

Implementation and Adaptation

So far, in most resource management committees around Mafungautsi Forest, collaborative monitoring systems have been successfully implemented and have been useful in joint learning, with stakeholders sharing the findings from their monitoring activities. The platforms for sharing the monitoring results that have been created range from community meetings (organized by communities, the councilors, traditional leaders, and committee members) to workshops organized and facilitated by the Forestry Commission. Some communities have started asking their committee members to provide progress and financial reports. The commission's officer is organizing regular meetings for the committees to share their monitoring and other experiences, reflect on them, and adapt their management strategies. In adapting their strategies, stakeholders in Mafungautsi have shown that collaborative monitoring is not an end but a means toward helping them realize their main objective—sustainable management of their resources.

Stakeholders originally expected that the monitoring subcommittees would take a leading role in organizing platforms to enable sharing of monitoring results and joint learning. However, the subcommittees have focused on ecological monitoring. The monitoring subcommittees had been created as auxiliary arms to the main committees, which were overwhelmed with work. As auxiliary arms, they had to report their findings to the committees, which later organized the platforms for people to share and learn from each other. Although it is important to define the roles of stakeholders in collaborative monitoring systems, these roles may change once implementation starts, and hence one has to be flexible.

Following several reports and complaints by communities on fund embezzlement by some resource management committee members, the Forestry Commission and the communities have embarked on monitoring the performance of their committees. In Chemwiro Masawi Resource Management Committee, the communities are demanding regular reports back to the constituents, rather than only upward reports to the Forestry Commission. Meetings are now held on the financial status of the committee. The Forestry Commission has also provided basic forms to help the committees write monthly reports. According to the commission's officer, this is an effective way of making sure that the committees report back to their communities. The officer also organizes workshops where information from monitoring activities is shared.

Lessons

One central lesson from our experiences to date is the importance of platforms for sharing information in any successful collaborative monitoring initiative. Creating these platforms requires the input of several stakeholders: communities, the resource management committees, the Forestry Commission, and traditional leaders. Where these platforms were not created, the monitoring systems have reverted to flawed past systems or disintegrated entirely.

A second lesson that we offer from Mafungautsi is that collaborative monitoring systems are dynamic, which requires flexible responsiveness from the implementers and facilitators. When developing the system, no one anticipated a need to monitor the performance of the resource management committees. However, after realizing the importance of collaborative monitoring processes, stakeholders started to monitor a much wider range of important issues in their lives, and not just forests.

Several challenges, however, have affected the ability of some systems to function. Some resource management committees still lack systematic documentation of monitoring results. At the inception of the collaborative monitoring exercise, a recordkeeping system was to ensure that all monitoring results were captured correctly and passed on to other stakeholders. Without properly documented information, wrong information can be fed back into the system, thereby giving wrong signals to decision makers and resource users alike. An example of a failed system is the Batanai Resource Management Committee.

Epilogue: Collapse of the System in Batanai

In the Batanai Resource Management Committee, the monitoring system failed because of political problems. Committee members from Batanai said they were approached by the ward councilor, who is a member of the ruling party, with a proposal to hand over some of the committee's money to sponsor a local party soccer team. The committee refused, saying it could donate the money only with the communities' approval and would have to consult with them about what was, after all, their money.

The ruling party officials were unhappy with this response, and it is alleged that they set the Batanai grass area on fire. Surprisingly enough, the suspected lead arsonist was the collaborative management subcommittee chair—also a staunch supporter and district chairperson of the ruling party.

Following these events, the communities decided to dissolve the monitoring subcommittee. So far, no one is taking the lead to monitor woodland resources. Some members of the main committee are still involved in monitoring but proceed according to the old, flawed system. The committee mainly monitors firewood collection, apparently to ensure that no one collects firewood without paying a new firewood levy. Without the monitoring subcommittee, the committee lacks information about the status of the forest resources.

CHAPTER 12

Learning to Monitor Political Processes for Fairness in Jambi, Indonesia

Trikurnianti Kusumanto

*I*N INDONESIA, WHEN ONE SPEAKS OF monitoring, the word *pengawasan* is often used; literally, it means "control." That this term is used not just by government agents in their programs and projects, but also by the staff of local nongovernment organizations, international development agencies, and donors, gives some indication as to how monitoring—in theory and in practice—is viewed by many in Indonesia. Activities are "controlled," and indicators tell planners and implementers to what extent progress is veering from planned targets. During project implementation, monitoring is often seen as an externally driven instrument to help project staff do their jobs. The so-called project beneficiaries hardly benefit from the monitoring activities—nor does anyone expect them to.

This chapter offers an experience in which monitoring was approached quite differently—as a means for local actors to learn collectively by keeping track of and influencing change processes. Our experience took place as part of participatory action research (PAR) in Jambi, central Sumatra, related to the adaptive collaborative management (ACM) of forests that involved local communities, two local partner NGOs (Yayasan Gita Buana and Pusat Studi Hukum dan Kebjikan Otonomi Daerah), and CIFOR (Figure 12-1). The team consisted of two CIFOR researchers (including the author) and four facilitators from the NGOs. The interactive research process took place between July 2000 and September 2002, but the experiences described here occurred over three months within that period.

The action research focused on two main issues: (1) building local governance structures and mechanisms that could support learning and communication for coordinated action, and (2) obtaining formal recognition of community rights to manage the village area and natural resources (Kusumanto et al. 2002a, 2002b). The first issue emphasized strengthening horizontal collaboration among the diverse community components. Given the current decentralization, limited government support to local people, and increasing socioeconomic stratification

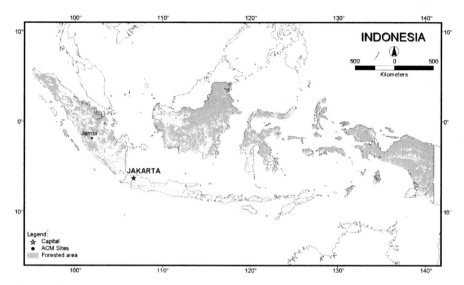

Figure 12-1. *Central Sumatra, Indonesia*

(between and within communities), our assumption was that devolution to local levels would be effective only if local communities could take on new responsibilities. The second issue was geared toward strengthening the communities' ability to negotiate on natural resources with external actors—neighboring communities and the district government.

Those two main issues were divided into six smaller issues, one of which, representation, is the focus of this chapter. Community stakeholders chose representation as a "learning topic" because they wanted to ensure that marginalized groups were not excluded from local governance opportunities and entitlement to benefits.

Our example involves local elections to form a new governance mechanism and could thus be considered "political learning." This is one of the four dimensions of social learning considered important in adaptive collaborative management, alongside knowledge sharing, communication and relationship building, and collective aspects of learning (Buck et al. 2001b). Although the learning focus revolved around choosing representatives in a local election, the process provided fertile ground for political learning in general. The type of learning-oriented monitoring that we describe below can help make forest governance accountable to those who depend on forest resources.

Context

The participatory action research was carried out in the Baru Pelepat Village of Jambi, situated on the upper part of the Pelepat River watershed. This watershed is part of the buffer zone of one of the four largest conservation areas in Southeast

Asia—the Kerinci National Park. We found considerable diversity of social groups, including the original inhabitants, settlers, nomadic groups, and customary elderly and other elite, and also a range of local institutions, such as customary institutions, formally appointed village government, and religious institutions. The community was characterized by weak social cohesion and poor communication and coordination between individuals, groups, and institutions. Decision-making related to local resource management was dominated by a handful of village elite with poor representation of the other groups.

The election of representatives to sit in a village decision-making body *(badan perwakilan desa)* was a good opportunity to learn what representation in decision-making entailed. Understanding this process and strengthening it would, we envisioned, enhance how the community managed its forest resources. Current decentralization policies require villages to set up their own councils. However, ineffective technical guidance from the government often led to the election of council members who already held local power or were favored by the district government. Inexperience in developing representation mechanisms at the community level thus tended to create the opportunity for village elite to capture the official right to represent the community.

When we began facilitating learning among the different stakeholders, we faced a complex set of contextual challenges:

- poor communication and social relations among social groups;
- weak coordination of natural resource management, mostly dictated by customary hierarchical structures and patron-client relationships;
- power imbalances, in particular between the village elite and other community groups, and between settlers and original inhabitants;
- unfamiliarity among many villagers with formal and structured ways of learning; and
- constraints on the participation of most actors—particularly the poor—in time-consuming learning activities because of their daily subsistence needs.

Creating Conditions for Local Monitoring

Confronted with the above challenges, we evolved toward a facilitation approach of the action research process that combined formal and informal methods, semi-structured and unstructured learning, and focused support to individual stakeholders as well as multistakeholder groups (cf. Kusumanto et al. 2002a, 2002b). In this way, we created an environment that enabled the local stakeholders to meet individual as well as collective learning needs. Our facilitation support was directed not only at the less powerful stakeholders in the community but also at those who held power, since we believed that learning should engage all relevant stakeholders, including the power holders.

In the action research process we facilitated, groups of stakeholders formed around issues they themselves had identified and then followed the iterative action research cycle of observe-plan-act-reflect. In this way, we aimed to create space for all groups to engage in focused deliberation, explicit learning, and adaptive action related to their concerns. Furthermore, the stakeholders would be encouraged to relate to each other individually and as a group to their natural environment. This would, we anticipated, create collective understanding of local issues.

However, the contextual challenges made us uncertain about what form of local monitoring would be appropriate and viable. We were also unclear about how we could involve local actors in developing such a monitoring system. Moreover, the participatory action research was not initially aimed at creating local monitoring mechanisms. Guided more by intuition than by reasoning, our team decided to start assisting local stakeholders in their learning processes by providing sufficient space for monitoring needs to emerge from within, rather than as externally driven constructs. Below is one example of this monitoring process.

Local Governance Mechanisms

To improve village governance, local actors needed to build understanding and awareness about notions of representation in community decision making. Stakeholder groups went through a process of experiential learning (creating knowledge through the transformation of experience; Kolb 1984) to choose their representatives to sit in the village council. The preparation for the election took three months, during which the experiences in this chapter occurred. As a team, we soon realized that the political nature of the process required us, as external facilitators, to help local stakeholders understand the unfolding power plays and changing political conditions. We were clearly focusing here on political learning (Buck et al. 2001b).

However, the skewed power relations in Baru Pelepat made it extremely difficult for the community stakeholders and our team to conceive of a monitoring system for the election process that would be transparent and accountable. At best, we felt we could create conditions for transparency and accountability in the way we facilitated and structured learning.

We focused on three complementary governance mechanisms: an election committee, community stakeholder meetings, and community-wide information dissemination. Of these mechanisms, the election committee appeared to be critical for the other two to be effective.

The committee, whose members came from each of the five hamlets, thus encouraging community-wide engagement, was given the mandate by all community groups to act on their behalf. Assisted by our team, the committee assumed several tasks: developing election procedures, registering eligible voters, disseminating information, fundraising as necessary, and preparing, implementing, and monitoring the election itself. In PAR terms, the activities carried out by the election committee can be considered as the action and observation components of the observe-plan-act-reflect cycle.

The second mechanism consisted of regular meetings among stakeholders organized by the committee to reflect on the actions that had been taken, report progress, consult about emerging issues, exchange information with the community, and make joint plans for follow-up action. Our team's role was to support the committee in facilitating communication among those participating at the meetings. This activity can be considered the reflection and planning components of the observe-plan-act-reflect cycle.

The third mechanism was an informal communication structure that emerged among villagers, in which the election committee members became key informants for those seeking information. Because the committee members represented all five hamlets of the village, information flows had a community-wide reach. Often, people spontaneously gathered at the committee members' homes to exchange information about the election. These informal channels of information (rather than those used during formal meetings) appeared to be important in the learning processes.

To our surprise, as will be discussed below, political process monitoring emerged as the election preparation progressed. People began to seek information through their interactions with one another, to assess situations that unfolded based on the information, and to organize themselves to deal with changing conditions.

Action and Reflection

The team's facilitation of the election committee and stakeholder meetings enabled participants to interact and communicate systematically. This appears to be the main contribution of an ACM-inspired process. Two aspects of learning seemed to foster collaborative and adaptive processes: learning from investigative action was more prevalent in the committee meetings, and learning from reflection manifested itself more in the stakeholder meetings.

Investigative action on problems with the coming election was carried out by the committee on behalf of the community. For example, when a powerful customary leader made attempts to covertly mobilize people who could help him seek political support, community stakeholders asked the election committee to investigate. Their findings were discussed in a stakeholder meeting, after which people made plans to counterbalance the customary leader's political maneuverings.

Learning from reflection in stakeholder meetings focused on the effects of actions and was instrumental for testing underlying assumptions that influenced people's acting and valuing. For instance, after collective reflections during a stakeholder meeting, original inhabitants halted their negative stereotyping of settlers. The reflections stimulated critical thinking about how the antagonistic behavior of original inhabitants affected the performance of a committee that also included settlers.

The contribution of ACM to the third mechanism, the spontaneously emerging local information networks, was less direct: the committee and stakeholder meetings fostered other and more diverse forms of sharing and related learning among community stakeholders.

Evolving Norms of Fairness

As the election approached, people were noticeably more concerned about whether processes were sufficiently transparent to give everyone a fair chance to participate. Most villagers could recall past abuses of power by community and customary leaders. The community had a long tradition of hierarchical control by local leaders, either via ethnic and family social networks, or via patron-client economic relationships. Many community members viewed the election as an opportunity to change this. In their collective monitoring efforts, people were motivated by one question: how to ensure that the election process was fair and power plays were minimized.

In socially divided settings such as Baru Pelepat and where socially or politically sensitive issues, such as the election of community representatives, are at stake, communication between political allies tends to take place behind closed doors and in informal encounters, often resulting in covert political arrangements. Community meetings rarely see frank debates, and decisions are often made with little concern for the politically weak or silent.

Nevertheless, this election experience in Baru Pelepat revealed that as election preparations progressed and stakeholders continued to share perspectives in and outside formal meetings, a collective view and awareness took shape as to how the election must be: fair with minimal power struggle. This collectiveness clearly manifested in people's actions, not just in their words.

Assessing the Learning

In evaluating the learning in Baru Pelepat, we first observed that local stakeholders developed a culture of curiosity: emerging social processes triggered by the election became the subject of scrutiny. Deliberations in the regular meetings and informal encounters outside the meetings served as a means for people to develop a capacity for inquiry. The more people exchanged information and perspectives, the more curious and alert they became, and the hungrier they were for further details.

Perhaps more significantly, they started questioning the context and consequences of changes, not just the changes themselves. For example, one change allowed women to participate in decision making. Women initially wanted to understand what their participation in community decision making entailed. Then they became curious about why men were now more willing to listen to them in meetings; what women's participation meant for themselves, their families, and the rest of the community; and what had previously hindered women's active engagement. Another example is offered by a customary leader who had held considerable power prior to the election but did not even pass the first election screening. He recognized that the past values of authority had to make way for new, more egalitarian values of governance and learned that wider changes demanded adaptation of long-established power structures.

Such attitudes of inquiry not only arose in facilitated formal and semistructured settings but continued in people's ordinary spheres of life. At the village grocery and in homes, people were involved in deliberations about the elections.

Increasingly Complex Questioning

The questions that people asked about monitoring a fair election also evolved, from relatively simple, technical questions at the beginning to more complex and politically value-laden questions as the election neared (Boxes 12-1 and 12-2). As

Box 12-1. Participatory Action Research Cycle

Observation: Community decision making does not involve all community stakeholders. In negotiation with "external" stakeholders about natural resources, views of community members are often poorly represented.

Phase 1

Plan: Hold community meeting to exchange views about participatory decision making and leadership. Explore ways to improve community decision making.

Action: Conduct meeting as planned, assisted by ACM team. Identify ways to improve stakeholder participation in decision making.

Reflection: Perceptions within community about leadership and decision making vary. Most think that old processes should make way for new, more democratic ones and that representation of all stakeholders is important. One way to improve representation is with village representative body. Such councils are regulated by district policy on village government. ACM team is willing to help set up council and elections. Community needs official support from district government for establishing council.

Phase 2

Plan: Approach district government to learn about its policy, administrative requirements, and what support it can provide for setting up council.

Action: Meet with district officials, with assistance from ACM team. Hold stakeholder meetings facilitated by ACM team to develop further understanding and exchange ideas about council.

Reflection: District government can provide administrative and financial support but has no experience in providing technical assistance because policy on village councils is new. ACM team is asked to facilitate. Community should submit proposed budget to district government and take four steps: form election committee, register eligible voters, nominate and screen candidates, then hold election.

Phase 3

Plan: Prepare for council election. Hold community meeting to discuss procedures for establishing election committee.

Action: Draw up overall election plan. Conduct stakeholder meeting and reach agreement on procedure for establishing election committee. Follow up with hamlet meetings to choose committee members from each hamlet.

Reflection: Committee members need certain skills and knowledge and many desire training. Amount of money required is now clear; proposal can be written and submitted to district government.

Phase 4

Plan: Train committee members in financial administration, budgeting, voter registration, election monitoring, and communication. Develop funding proposal.

Action: Conduct training for committee members. Make election budget and write funding proposal. Disseminate election information to community.

(continued)

Box 12-1. *(continued)*

Reflection: Election committee is not supported by village leaders and must handle every-thing on its own. External assistant (from ACM team) is needed. Bureaucratic require-ments demand time, effort, and travel to government office.

Phase 5

Plan: Seek commitment from village leaders to support election process. Register eligible voters.

Action: Hold meetings with election committee and village leaders to improve communi-cation and encourage cooperation. Conduct voter registration in five hamlets.

Reflection: Village leaders are more willing to cooperate and show appreciation for com-mittee's work. Registration is more complicated than expected because population data are not up-to-date or complete.

Phase 6

Plan: Finish voter registration. Develop criteria for eligibility of candidates and screen potential council members.

Action: Finish registration. Hold stakeholder meeting to determine criteria for eligibility and screen candidates.

Reflection: Bad population data complicated election preparations. Collaborative devel-opment of screening criteria has made all stakeholders feel responsible for monitoring the screening process. Political maneuving of certain community members should be monitored.

Phase 7

Plan: Intensify monitoring of election preparations and political maneuvering. Increase communication about election processes. Conduct and monitor election.

Action: Monitor election preparations. Hold information dissemination activities. Conduct and monitor election.

Reflection: Dissemination of information encouraged community to participate in moni-toring and ensured transparent processes. Hours of polling seemed appropriate for work-ers' schedules.

Phase 8

Plan: Count ballots and announce election results.

Action: Count ballots and announce election results.

Reflection: No significant problems encountered.

---

people gained new insights about equity in elections and what issues needed to be monitored, the monitoring processes themselves were adjusted to fit the increas-ingly complex issues.

Although our facilitation of learning processes through the committee and stakeholder meeting mechanisms helped focus and structure the learning, it was people's own efforts to understand the unfolding social processes, generated by

---

Box 12-2. From Technical to Political Questions

One question that stakeholders first regarded as critical was how to ensure that villagers actually voted in the district to which they were assigned—a technical, administrative question. Some villagers and their families had homesteads in two administrative districts and could hence mistakenly or, as some feared, deliberately register twice for the election. Village population data were neither well organized nor complete, creating room for suspicious administrative practices. A later question revolved around the screening of eligible candidates—a question with political overtones and not a simple administrative matter.

---

the coming election, that directed their learning and therefore their monitoring of their village election.

Power Relations

A third learning aspect that we observed in the field concerns power relations. As noted above, we assessed that an open election monitoring system would not be viable in the sociopolitical setting in Baru Pelepat. Thus far, the social learning generated by the three complementary monitoring mechanisms—an election committee, the stakeholder meetings, and the information networks—had supported collective monitoring and a culture of inquiry. Could monitoring also deal with power differences?

We observed that stakeholders' information analysis was able to counter attempts at power abuse. The group discussions in the community meetings went beyond merely sharing information or views or bringing together pieces of knowledge to create a larger shared construct; rather, stakeholders gradually built a collective experience by analyzing situations and identifying and correcting mistakes (Box 12-3). In this way, checks and balances became part of the analysis.

Besides analyzing concerns, participants of the sessions we facilitated also started discussing issues that they previously would have considered too socially sensitive for public debate. Linking people's experience to a process of joint analysis—that is, taking people's concrete realities to a level of abstraction—appeared to have made sensitive issues discussable and individual mistakes recognizable. Our facilitation style emphasized debating about substance rather than relational issues, and that approach seemed to make difficult and potentially divisive issues discussable. For example, ineffective leadership was discussed by focusing debates on how such leadership affected peoples' livelihoods, rather than considering actual personal relationships.

During these discussions, people raised an old issue of power abuse by a corrupt former village headman and other elite during implementation of a past government resettlement project in Baru Pelepat. Project funds and materials had been misused. Encouraged by the idea that the coming election could make governance more accountable, villagers pursued this example to determine what could be done to avoid similar power abuses.

Box 12-3. Built-In Corrective Action

Stakeholder meetings organized by the election committee included discussions on such concepts as leadership and community representation. Local residents noted that leadership had been weak and attributed it to the inability of most leaders to listen to the people they were supposed to represent, focusing only on personal interests, and to the absence of mechanisms that could check power abuses. Although this conclusion may seem straightforward, it needed thorough collective analysis based on people's different experiences with the lack of leadership and lack of representation of their interests.

We focused group discussions on building collective understanding about the underlying causes of corruption based on people's own experience, as well as on how abuses affected people's lives and the community as a whole. By putting people's experience about corruption in a wider context, in terms of both causes and effects, people had a clearer idea about how to build in checks and balances and were better prepared to make and accept suggestions for corrective action.

Lessons

Our experience in Jambi describes how monitoring can be internally driven and supported by external facilitation and benefit local processes of social change. The discussions that took place among the stakeholders motivated them to begin monitoring, which in turn encouraged more learning, and to institutionalize these monitoring mechanisms.

As stakeholders' capacities evolved, monitoring needs developed in parallel, monitoring questions changed, and therefore also the types of information being sought and analyzed. As conditions in Baru Pelepat increasingly supported transparent and accountable social interactions, monitoring emerged from within: people developed a capacity for inquiry to support their own coordinated action. Checks and balances of power plays were inserted into processes of learning and inquiry.

We saw that monitoring and learning are clearly interconnected and that different social forums are required for monitoring to engage the entire community. Our field experience also showed us that monitoring can go beyond a technical exercise for tracking implementation of development projects and be applied to local governance.

Now, two years since these events unrolled, the political learning during the election activity described here has clearly left its trace. Natural resource management appears to be more accountable to diverse community groups and more inclusive of their perspectives. Recently, for example, an oil palm company sought to coopt the village headman to get a permit for forest clearance and begin an oil palm plantation, but failed. The majority of community stakeholders did not welcome the large-scale operation. Had a similar situation presented itself in the past, community representatives might have acted out of personal interest, without considering the views of their constituencies, and external actors could likely have tapped into the power of the community elite.

Our experience implies that in Indonesia, where most participatory approaches show limited success in equitable distribution of forest benefits, internally driven monitoring can serve as a basis for political learning. Many forestry projects systematically exclude less powerful groups from decision making about forest benefits or are captured by the forestry agency or the local elite. Linking monitoring to learning, on the other hand, offers potential advances in accountable forest governance.

PART VI

Conclusions

CHAPTER 13

Expanding Views about Collaborative Monitoring

Irene Guijt

MONITORING HAS, TO DATE, BEEN SEEN quite narrowly by those work-ing in natural resource initiatives for sustainability. Development project managers mainly consider monitoring a tool for controlling proper use of money and determining whether work is progressing as planned (Roche 1999; Woodhill 2004). Researchers generally view it as a data collection process, although some are stretching its use to include a more analytical dimension and viewing resource management (and related policy processes) as ongoing experiments (Lee 1993; Bosch et al. 1996; Probst 2002). Evaluation specialists in ex ante environmental and social impact assessment are required to develop monitoring plans (IAIA 2003), but often the quality is poor, emphasis is on compliance monitoring, and monitor-ing for learning falls outside assessment practice. In all cases, monitoring has been seen as a fairly mechanistic exercise comprising little more than indicator selection and data collection.

This volume on monitoring for adaptive collaborative management (ACM) fun-damentally challenges that perspective and offers a new perspective. It considers monitoring a process of socially negotiated learning. The contributors stress the need to negotiate core definitions and methodologies for monitoring, as well as new understandings about resource management that emerge from joint information analysis. This expanded view of monitoring allows it to become a more versatile concept that better reflects on-the-ground realities. Indicators make way for ques-tions and issues, data collection makes way for forums and debates, and single defini-tions make way for shared understandings. The experiences in this volume contrib-ute to the ongoing discussion about making science truly collaborative and serving those on the margins (Rocheleau 2003; Leach et al. 2005; Buhler et al. 2002).

The existing literature has tended to consider learning an inevitable by-prod-uct of a data-focused monitoring system. This book shows that learning does not happen through imposed concepts and methods; they do not capture the interest of those involved and are quickly abandoned. Instead, a monitoring program that seeks to institutionalize learning among local stakeholders must be based on clarity of choices and realistic expectations.

Amid the diversity illustrated by the case studies, patterns of inspiring ideas and cautionary signs can be discerned. This chapter summarizes the lessons from monitoring practice and, in so doing, offers ideas that can help those embarking on or already involved in monitoring initiatives. First, it summarizes the benefits that are emerging from the experiences in this book, offering a grounded set of expectations instead of the commonplace generic hopes. It then identifies the main obstacles and questions that the contributors have had to deal with during design and implementation. Conceptual and methodological issues are found to need more explicit attention, particularly in the startup phase, and a good understanding of stakeholders and their processes is important. The final set of lessons concerns capacity building and phasing out external facilitation. The chapter ends by reflecting on when monitoring can add value to resource management and the criteria that can be used to recognize monitoring that triggers learning in that context.

Appreciating the Benefits

The frank accounts in this volume provide a refreshing view on monitoring. They describe the sometimes considerable difficulties encountered in creating sustained monitoring. One lesson is that capacity building requires time and resources on all sides—it does, after all, concern a relatively novel process of systematic and focused tracking of resource issues by local groups. But collaborative monitoring is not only about laborious efforts, doubts, and dilemmas.

Each experience also shows that monitoring can provide benefits—and that these are more diverse than are usually assumed (Table 13-1). The benefits may not emerge in all cases, nor are they needed in all situations. Furthermore, the chapters do not establish unequivocally that these effects are sustained. Table 13-1 lists only those benefits that are evident from these experiences; others may emerge in different contexts and applications. Furthermore, many of these monitoring experiences are in their initial stages, and additional benefits may ensue later.

In almost all cases described in this book, working together on a monitoring initiative has helped those involved understand more about institutional, environmental, and resource problems. For example, in Zimbabwe (Chapter 6), information on the volume and quality of broom grass led to the identification of limits on its use. In Malawi (Chapter 10), Kamoto writes, "The clear benefits accruing from this nontimber forest product [beekeeping] stimulated a wider community interest and willingness to learn about the resource base."

This relates to the expectation that collaborative monitoring can improve mutual understanding of forest management visions and options. In Nepal (Chapter 4), a simple exchange of views on the indicator "improved forest condition" showed that a farmer defined it as "good ground cover with grass," while the forest guard defined it as "forest cover where we can't see people moving inside the forest." Articulating such differences forms the basis for constructing new shared visions. Mutimukuru's experience in Zimbabwe (Chapter 11) led to a new understanding of the role of the forest guard, and in Malawi (Chapter 10), goals for sustainable beekeeping could be set for the forest reserve. As stakeholders constructed their monitoring plans and shared their ideas about what forests meant, where problems occurred, and what might be needed, greater collective clarity appeared to emerge.

Joint monitoring has also enabled more effective decision making about forest management—either more informed decision making, as in Zimbabwe and Malawi (Chapters 6 and 10), or more equitable decision making, as in Nepal and Cameroon (Chapters 8 and 9). This is the basis for reducing conflicts over resources. Cronkleton explains that the choice to focus on monitoring the distribution of benefits was made because of the potential for conflicts, given the limited understanding by community members about the decisions and distribution process.

Critical for adaptive management of forest resources is the ability to look at one's resource base with fresh eyes. Colfer (2005a) discusses the importance of

Table 13.1. *Reported Local Benefits of Collaborative Monitoring*

Type of benefit	Source (chapter in this volume)
Improved understanding of institutional, environmental, and resource problems	Pokorny et al. (2); Santos et al. (3); Cronkleton et al. (5); Nyirenda and Kozanayi (6); Hartanto (7); Oyono et al. (9); Kamoto (10); Mutimukuru et al. (11); Kusumanto (12)
Mutual understanding of forest management visions and options	Pokorny et al. (2); Santos et al. (3); Paudel and Ojha (4); Cronkleton et al. (5); Nyirenda and Kozanayi (6); Hartanto (7); McDougall et al. (8); Oyono et al. (9); Kamoto (10); Mutimukuru et al. (11)
More informed and/or equitable decision making about forest management	Paudel and Ojha (4); Cronkleton et al. (5); Nyirenda and Kozanayi (6); Hartanto (7); McDougall et al. (8); Oyono et al. (9); Kamoto (10); Kusumanto (12)
Increased capacity and willingness to question previously accepted norms (institutionally and technically)	Cronkleton et al. (5); Nyirenda and Kozanayi (6); McDougall et al. (8); Kamoto (10); Mutimukuru et al. (11); Kusumanto (12)
Resolution or management of conflicts	Paudel and Ojha (4); Cronkleton et al. (5); Nyirenda and Kozanayi (6); Oyono et al. (9); Kamoto (10); Mutimukuru et al. (11)
Shift in perception from monitoring as policing to monitoring as local benefit	Santos et al. (3); Paudel and Ojha (4); Kamoto (10); Mutimukuru et al. (11); Kusumanto (12)
Higher quality of social and organizational interactions (social capital), communication, and (inter)group skills	Pokorny et al. (2); Santos et al. (3); Paudel and Ojha (4); Nyirenda and Kozanayi (6); Hartanto (7); McDougall et al. (8); Oyono et al. (9); Kamoto (10); Mutimukuru et al. (11); Kusumanto (12)
Increased equity in who is heard and who benefits	Cronkleton et al. (5); Nyirenda and Kozanayi (6); McDougall et al. (8); Kamoto (10)
More sustainable forest management practices, fewer harmful forest resource practices	Santos et al. (3); Cronkleton et al. (5); Nyirenda and Kozanayi (6); Hartanto (7); McDougall et al. (8); Kamoto (10)

allowing surprise to guide resource management. But being open to surprise requires being willing to question institutional and technical norms. Take the case of Zimbabwe (Chapter 11), where much time was spent creating new terms of reference that defined roles and responsibilities for all resource users and managers. Letting go of old ways of seeing resources use and management also extends to the notion of monitoring itself. In several examples, forest users and government forest guards had to become accustomed to the idea that monitoring was no longer about policing but now meant collective guardianship.

Such adjustments can help improve social and organizational interactions. In Zimbabwe (Chapter 6), tension between the resource management committee and the broom grass group dropped, as did tension among grass harvesters arising from previously unfair access to grass spots. One example of improved local forest governance comes from Nepal (Chapter 8), where the forest user group committee's meetings were not open to general members, and few people knew about its decisions. Now input is actively sought and provided through shared agenda-setting processes—and decisions are fed back to the hamlet level. The villagers in Jambi, Indonesia (Chapter 12), found themselves on a steep learning curve as they set out to ensure fair election processes. The construction of a representative election committee enabled community-wide dissemination of information, and villagers actively engaged in both formal and informal meetings.

Finally, of course, successful forest use means that harmful resource practices are reduced and benefits are shared equitably. Kamoto's account from Malawi (Chapter 10) focuses on how indicators gave the community forest patrols a list of things to look for during their rounds and helped provide information that was subsequently used to reduce theft from beehives. In the Philippines (Chapter 7), Hartanto tells us how recording illegal activities helped cooperative members realize that they did not know how to report such incidents to authorities in ways that would elicit a quick response—and they set about filling this gap.

The articles collectively suggest that collaborative monitoring represents a process that can help institutionalize new norms for resource use. These norms include making equity a reality, dealing with (rather than avoiding) conflicts, and shifting entrenched and erroneous perspectives on what is "good" forest practice. There is indeed much to value in collaborative monitoring—it is not wishful thinking.

However, collaborative monitoring can entail considerable changes in the status quo and thus some (tacit) agreement on the desirability of new norms is required among the main stakeholders. Without this initial agreement, the subsequent change process may well pose insurmountable challenges for any facilitators. Therefore, a good dose of realism is also required—as borne out by all the experiences. The remainder of this chapter discusses some of the most critical considerations for collaborative monitoring to be effective.

Clarifying Concepts and Process

This book has repeatedly referred to ideas that some may consider vague—learning (and social learning, at that), collective analysis, critical reflection, participatory action research, and so forth. Certain other terms repeated through the book are often associated with rigidity—monitoring, criteria and indicators framework, and objective hierarchies. For many engaged in local forest management, such concepts and processes can reek of past problems or may be entirely unfamiliar. Complete clarity on everything is not possible; heed the warning that the person who insists on seeing with perfect clearness before deciding never decides. Nevertheless, seeking clarity on purpose and concepts can avoid problems. Clarity is also needed about the starting point of discussion, whether it is an existing forest management plan (even one of poor quality), a research question, or a policy gap. Finally, conscious decisions are needed about how to update the monitoring processes so as to build in the learning cycle.

Being Clear about Purpose

The importance of purpose has been discussed but cannot be emphasized enough. As a Zen saying goes, "clarity of purpose, clarity of understanding." Once it is clear what is needed, then the basis for subsequent decisions becomes clear. Chapter 1 gives a range of possible purposes that can drive a collaborative monitoring system.

Ensuring accountability for funding agencies and learning for improvement are perhaps the most common purposes. Monitoring is also essential for those who advocate sustainable forest management practices: tracking nontimber forest product harvesting may help show what volume can be extracted without adverse consequences, and tracking fires and other resource damage can trigger corrective action. Monitoring to ensure legal compliance and private sector voluntary commitments (Global Forest Watch n.d.) is one concrete example of advocacy-oriented monitoring that is currently receiving considerable attention.

Initiatives that take a monolithic approach to collaborative monitoring by squeezing all information needs and analysis into a uniform system cannot accommodate a diversity of learning strategies. Early on in designing a collaborative monitoring initiative, stakeholders need to clarify the main purpose and determine whether multiple purposes need to be met (Table 13-2). If stakeholder composition changes, the purpose may need to be revisited so that all those involved share the monitoring aims.

The purpose strongly shapes implementation, affecting the time frame and the extent and nature of stakeholder participation. Being clear about purpose helps make the monitoring operational, since those involved can then determine the time frame and how it links to decision making, their participation, and the depth of analysis and rigor required (Guijt forthcoming). It also gives stakeholders the flexibility to develop separate and complementary monitoring processes to fulfill different purposes; for example, one group can focus on local decision making while others address the information needs of external stakeholders.

Time frame. Should efforts be invested in establishing long-term monitoring mechanisms, or is short-term tracking sufficient? A short time frame may be adequate for gathering information about a known phenomenon to present to a particular group (purpose 7, Table 13-2). However, longer-term monitoring

Table 13.2. *Differentiating Monitoring Types According to Purpose*

Monitoring as process of...	Core purpose	Example
1. Financial accountability	Maintain funding	Standard reporting of expenditures, activities, and outcomes (possibly impacts) effects
2. Strategic reflection	Examine strategy and test underlying assumptions	Tracking changes in livelihood levels among case study families to assess validity of assumption that ACM leads to sustainable livelihoods
3. Capacity strengthening	Improve individual or organizational performance	Reviewing colleagues' and own performance to improve implementation process
4. Tracking context	Keep up-to-date on context of implementation	Reporting on state of the environment
5. Research	Examine uncertainties and formulate new questions	Using indicators to monitor and improve agroforestry systems through participatory research with farmers
6. Ensuring transparency and trust	Maintain transparency in use of resources	Reporting by forest users on who has harvested what, where, when, and how
7. Building critical mass of support for concern or experience	Sensitize wider social group to gain support for joint action	Tracking selected families (by themselves) for phosphorescence-tinted pesticides to gain awareness of need for collective action to seek alternatives (Stephen Sherwood.)
8. Public policy advocacy	Push for policy change	Tracking infringements of forestry laws or forestry concession plans to argue for public policy change

Source: Guijt, forthcoming.

mechanisms are needed for ongoing state-of-the-environment reporting (purpose 4). Ensuring that these correspond with local rhythms is important. Santos and her colleagues in Brazil (Chapter 3) noted the slow and fragmented rhythm of monitoring efforts that resulted from existing community decision-making processes.

Link to decision making. Who needs the information, who analyzes it, and which decision-making processes will the findings influence? If monitoring is to inform decisions, the decision makers' priorities, processes, and schedules are paramount. Take the case of purpose 6 (Table 13–2)—monitoring for transparency and as the basis for trust, of which the Bolivia, Malawi, and Zimbabwe cases are good examples (Chapters 5, 6, and 10). The information that is collected and analyzed must be relevant for the group that handles transgressions. This contrasts with purpose 2—strategic reflection—which must involve several stakeholder groups, each with its own processes and calendar of activities. Joint monitoring requires more inclusion and, above all, clarity about how the final decisions will be made.

Degree of participation of stakeholders. Who must be involved? What can be gained by including different groups—local ownership through joint analysis, agreement on visions for a forest, acceptance of decisions made, or something else? What should each group's role be? Regular strategic reflections probably require more widespread participation than state-of-the-environment reporting, which is often the province of trained scientists (purpose 4, Table 13–2); although community monitoring that feeds scientific reporting is increasingly common.

Degree of rigor. Several contributors commented on the unnecessary detail imposed by the criteria and indicators framework (Chapters 2, 3, and 7). Seeking comprehensiveness to ensure sufficient rigor can lead to stagnation or abandonment of efforts. But then, what is sufficient rigor—whose standards count? Mobilizing people (purpose 7, Table 13–2) may require less scientifically rigorous data than research (purpose 5), but auditing (purpose 1) must comply with certain standards of rigor. The rigor question requires distinguishing between data that one would like to know and those that one *needs* to know.'

Agreeing on Core Concepts

"The chief virtue that language can have is clearness," Hippocrates said, "and nothing detracts from it so much as the use of unfamiliar words." The core concepts of collaborative monitoring—participation, social learning, monitoring, communication, critical reflection—are ambiguous terms that are often used in differing and normative ways. Different stakeholders may well hold different views on what participation means in practice or how to balance reflection with data collection. Discussions about these core concepts are important to build a shared understanding or at least identify where differences may lie.

Early in an ACM process, terms like *monitoring, indicator,* and *learning* need explicit discussion. Paudel and Ojha (Chapter 4) found much confusion about *indicator,* as did Santos and her colleagues in Brazil (Chapter 3). Chapter 11 describes how forest guards and villagers jointly redefined the notion of monitoring early in their process, since the existing understanding of monitoring as a policing function was inhibiting the potential for learning. Discussions in Zimbabwe gave Mutimukuru and her colleagues an opportunity to reflect with resource users on their monitoring system and also the importance of collaborative monitoring for adding value.

Particularly important is clarity about *learning.* ACM initiatives seek information-driven adaptation as a continual self-correcting and improving process. This is a very different kind of learning—more dynamic, reflective and experience-based—than taking courses or obtaining information. Several chapters in this volume make it clear that data collection alone does not lead to learning (Chapters 5, 6, 7, and 10); it is the insertion of data into discussions and decision making that leads to action-oriented improvements

Knowing Where to Start

So what forms the starting point for discussions on monitoring? Is it the existing local forestry arrangement or an external plan of action for the forest area or the research question? If monitoring is to improve financial accountability, strategic reflection, and capacity strengthening (Table 13-2), then the most logical starting point is the forest management plan. This might be a funding proposal with a hierarchy of objectives or an agreed upon joint management plan. Such plans can be a useful basis for comparing planned with actual performance.

Together with the stakeholders, one can assess whether a clear and accepted forest management plan exists on which to base the monitoring system. If not, before the monitoring process can be designed, time must be spent formulating a plan or rewriting unclear goals, objectives, and assumptions. Kamoto describes how in Chimaliro, the forest management plan written by foresters had no local ownership and was not being used to guide improvements. She undertook a more participatory planning exercise that became the basis of subsequent monitoring work (Chapter 10).

Planning is commonly separated from monitoring, but they are part and parcel of the same process. Therefore, it is not surprising that Paudel and Ojha (Chapter 4) included identification of priority issues and planning as essential parts of an integrated learning process on forest management. This was echoed in Brazil (Chapters 2 and 3). And Hartanto's experience in the Philippines (Chapter 7) involved identifying priority areas for developing and improving plans.

Other purposes call for different starting points. For example, if tracking (purpose 4, Table 13-2) is needed, the monitoring work will focus on developing a process for a core set of environmental indicators, such as for the state-of-the-environment reports produced by some countries. If monitoring is for research (purpose 5), then the starting point will be the variables relating to the research question. If monitoring is to be used for public policy advocacy (purpose 8), then the starting point will be problematic policies that can be critiqued by using monitoring data or policy improvements that need to be supported with monitoring data.

Completing the Learning Cycle

"He who learns but does not think is lost. He who thinks but does not learn is in great danger." Confucius articulates what the ACM work is all about—learning as a necessary process if forests are to survive and forest communities are to have dignified livelihoods and well-being.

This book is replete with discussions on the importance of linking information and analysis. In Chapter 1, Figure 1-2 shows that monitoring data must be accompanied by reflection to lead to application of lessons and insights. Building the full learning cycle into the process takes monitoring beyond simple reporting (Figure 13-1). ACM initiatives must demonstrate what happened, the quality of the monitoring, and the immediate results. Learning that leads to action must involve analyzing why changes occurred and what the next steps might be (Figure 13-1). As

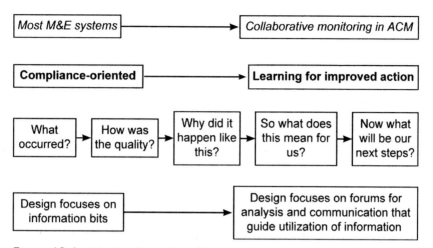

Figure 13-1. *Moving from Compliance to Learning in Monitoring*

Paudel and Ojha (Chapter 4) concluded, "Monitoring remained at the center, not as a rigid and hierarchical framework of indicators, but as a dynamic link between learning and planning."

Kamoto's example (Chapter 10) of the evolution of monitoring in the Chimaliro forest is rich in detail. She recounts how the patrols took place (what happened) but that the patrols hesitated to point fingers at culprits (quality of monitoring), so honey thefts continued (immediate result). Reflections by the coordinating committee about the problem and what needed changing led to new steps that ensured that sanctions were imposed, and honey thefts decreased. In Bolivia and Nepal (Chapters 5 and 8) the difference made by critical reflection is equally evident. And Kusumanto (Chapter 12) illustrates how repeated cycles of critical reflection on political processes allowed more complex questions about accountability to emerge. A critical component in all cases was joint analysis of the information and debate in accepted forums with decision-making power.

Considering Information Quality

Notwithstanding the centrality of learning, the quality of data must not suffer. As Ottke et al. (2000, 1) say, "We emphasise communication skills because the best data in the world is useless if it lies fallow. But credible data is also critical to convince skeptics and engender trust."

Solid information is critical—without credibility, evidence that has been gathered will not influence any decision maker. However, often a disproportionate amount of time is invested in trying to reach agreement about "the "best" indicators, at the expense of developing discussion and decision-making forums where information is analyzed and communicated and the learning cycle can come full circle.

The examples in this book suggest that ensuring that the required information is good enough (rather than perfect) is as important as making sure that the discussion and decision-making forums are appropriate and effective. Information and analysis must be in balance.

The experiences show that both qualitative and quantitative information is useful. The choice depends on the question being asked and the user of the information. In Dimako, Cameroon (Chapter 9), the monitoring of land transfers to two villages focused mainly on process issues, such as how the council was returning the reallocated agricultural land and how the two villages were sharing it. But they needed quantitative data in the form of social and agroecological maps for verification. In Nepal (Chapter 8), heterogeneity analysis helped the forest user group to "track and assess who was participating, contributing, and benefiting . . . and trends in participation, contributions, and benefits for marginalized versus non-marginalized members proved critical . . ." Thus quantitative data enabled both a quantitative and a qualitative analysis of equity.

Many monitoring information systems fail because they cannot explain why something is occurring. A useful distinction can be made between contextual and noncontextual methods of data collection (Booth et al. 1998, cited in McGee 2004): "methods which are contextual . . . are those that attempt to capture a social phenomenon within its social, economic and cultural context . . . On the other hand, any large-scale household survey . . . will lie at the non-contextual end of this spectrum." One must decide whether contextual insights are needed to explain forest quality or use, and that may determine what kind of information is needed.

Information needs inevitably change over time—many indicators selected in the beginning may lose significance as participants become more adept at identifying what information is relevant for their resource management activities. Kusumanto (Chapter 12) describes how citizens' increasingly complex questions during the preelection process inspired them to seek out different types of information. But not all information needs have a short shelf life. Some data remain important, such as the use of specific forest resources. Most monitoring efforts therefore consist of a mix of long-term fixed indicators and a set of shifting information needs.

Understanding Stakeholders and Their Processes

It requires considerable political astuteness to navigate the human relationships within a forest area. The contributors have highlighted four important considerations. First is getting agreement on how inclusive to make the monitoring design and implementation. A second consideration relates to the forums that are needed for sharing, debating, and decision making. A third aspect is choosing an organizational affiliation and entry point for the work. A final consideration relates to the transaction costs for stakeholders in any participatory process, including collaborative monitoring.

Agreeing on Roles and Extent of Participation

Critical to all the experiences is the decision about how inclusive to make the monitoring process. This decision is needed prior to designing the process (who should help construct it?) but also in implementing it (who is needed to make it work?). Opening participation to all possible stakeholder groups may be desirable to engender shared ownership, but as Santos and her colleagues discovered in Brazil (Chapter 3), the range of perspectives may be too divergent to allow for convergence around a common monitoring agreement. Accordingly, they decided to focus initially on the stakeholder group of central importance, the rubber tappers who were the main target group for the new benefit-sharing forestry practice. In the Philippines (Chapter 7), all stakeholders negotiated a shared monitoring framework and then allocated responsibilities based on formal mandate and interest. Kusumanto (Chapter 12) discusses how critical it was for the facilitators to adopt a socially inclusive approach to monitoring the preelection process in Jambi, since norms of fairness needed to shift across all stakeholder groups.

Deciding which stakeholder groups to include and what is expected of each group requires careful thought. The desire for local ownership must be balanced with considerations of whether stakeholders have the capacity to contribute in a meaningful manner and are willing to be involved. In Zimbabwe (Chapter 11), the terms of reference for the monitoring initiative included specifics on how the monitoring subcommittees should link with other stakeholders and organize forums for stakeholders to analyze the information. As a result, the roles of all other stakeholder groups were also clarified.

Roles need regular redefinition. As an ACM initiative shifts and capacities grow, new configurations of stakeholders may become feasible. Particular care must be taken to avoid token participation, which can turn the good intentions of learning and improvement into a technical exercise in data collection. In Nepal (Chapter 8), the forest user group shifted from a passive group dominated by the chairperson to an active committee driven by a self-monitoring process with a renewed constitution and operational plans; the change occurred because their participation ceased to be token.

Selecting Forums

Monitoring processes include identifying where information is to be shared and analyzed. Possibilities include general community meetings (Chapter 11), special forestry management group meetings (Chapter 8), and specific resource use committee meetings (Chapter 6). As mentioned in Chapter 1, such forums or spaces can be closed, invited, or created (Cornwall 2002). "Spaces," as used here, comprise those "opportunities, moments and channels where citizens can act to potentially affect policies, discourses, decisions and relationships which affect their lives and interests . . . Power relations help to shape the boundaries of participatory spaces, what is possible within them, and who may enter, with which identities, discourses and interests" (Gaventa 2006, 5–6).

In closed spaces, the decision-making group excludes others. Such spaces, typically composed of elected representatives, may be effective but are often places where decisions are made by a select few. In Nepal (Chapter 8), for example, "annual plans and activities were generally made and carried out at the discretion of the chairperson of the executive committee." Decisions made within such "closed spaces" should follow wider consultation and then be shared. Sometimes, however, closed spaces can be very useful and need to be created. In Malawi (Chapter 10), a coordinating committee was formed to oversee issues from three forest areas. This committee consisted of representatives from the three areas, and its decisions—behind closed doors—circumvented reluctance to arrest honey thieves.

Created spaces can be the domain of those who are commonly marginalized. Although such spaces can allow for discussions and decisions on people's own terms (and not on the authorities' terms), they can also suffer from prejudices and biases related to gender, age, poverty, ethnicity, or caste. McDougall and her colleagues (Chapter 8) give a poignant example of a community-based forest user group whose decision making and benefit sharing were dominated by the local elite while low-income and low-caste people and women were marginalized. Their ACM efforts included changing the decision-making and planning processes to ensure greater opportunity for marginalized group members.

Sometimes an existing social or administrative unit can become a created space, offering opportunities for learning that do not exist elsewhere. Both experiences from Nepal (Chapters 4 and 8) explain how shifting the attention from the village level to the hamlet level allowed more forest users to share information and reflect critically on progress. At other times, resource management committees that mix marginalized groups with other groups are formed for specific purposes and take on crucial roles. In Cameroon (Chapter 9), for example, an ad hoc committee of local and external representatives was created to follow up initial ACM work and became an important space for creating and implementing mechanisms for sustaining change.

Between the closed-door forums and the open-access opportunities are the "invited" spaces (Cornwall 2002), whose benefits for marginalized stakeholders can be difficult to assess. Hartanto (Chapter 7) describes an interesting case in which a monitoring effort was undertaken by different stakeholders who agreed to work as genuine equals, but ultimately preexisting power differences prevailed. In her conclusion, she notes that the powerful state players had helped design the monitoring process but were then hesitant to support the local organization; "a jointly developed framework does not necessarily mean shared responsibility for undertaking the work involved." The local government agency had a plan that would force all local organizations to carry out a different environmental performance monitoring system, which would discourage local monitoring and related ACM efforts. Thus Hartanto observes that more powerful stakeholders need to provide a genuine opportunity for collaborative monitoring to be based on learning and not policing. Such opportunities need to be constructed—they are rarely given freely.

Finding the Entry Point

A thoughtful entry strategy includes consideration of the organizational affiliation of the facilitators, the local power dynamics between organizations involved in the resource management, and the level where monitoring needs to be set up and sustained.

For facilitators, alliance with a local organization may smooth the entry. Dos Santos (Chapter 3) facilitated a process in Brazil while she was the staff member of an organization not involved in the local community. This neutrality gave her credibility and trust. When she later joined the staff of an NGO with a stake in the outcomes, her neutrality was called into question (but eventually confirmed). If her initial entry had been through this group, she might not have won the community's trust.

The lesson of Mutimukuru and her colleagues (Chapter 11) is powerful. Before initiating their collaborative monitoring process, they spent considerable time thinking about the best entry point. The context studies they undertook gave them insights into local power dynamics and helped them select the Forestry Commission as the best option because of the support it could offer. Sometimes, a neutral entry point is best. Oyono and his colleagues (Chapter 9) describe how the ACM team in Cameroon convened a meeting of different stakeholders and maintained an unaligned position to ensure fairness and facilitate access to all the parties involved in a land conflict.

Getting a feel for the organizational and social landscape takes time and political perceptiveness. Monitoring can get off to a wrong start if the facilitators are unclear about where to focus their efforts. Paudel and Ojha (Chapter 4) began by looking to the formal forest user group as the vehicle for monitoring ACM but quickly stepped back and found more suitable social processes and structures. They worked on strengthening the village level as the backbone of the monitoring process, as did McDougall and her colleagues (Chapter 8).

Recognizing Transaction Costs

Participation is not free. Forest guards, facilitators, and researchers are paid to undertake monitoring work, but for others—usually the forest users—time spent in meetings and debates means less time earning an income. The transaction costs for local people and organizations are considerable (Table 13–3). Yet they are rarely funded.

Table 13.3. *Transaction Costs for Collaborative Monitoring*

Time spent in collaborative monitoring	Costs
Interacting in interviews, focus groups, committee meetings, workshops, seminars, community meetings	Time for other activities, which may be productive, reproductive, developmental, political
Analyzing data, opinions, problems during mapping, transect walks, computerized data analysis	
Sharing information through theater, dance, storytelling, video and radio productions	Cash outlays for collective activities, such as joint meals, food, transportation, accommodation
Traveling to and from meetings, data collection sites	Social position in relation to nonparticipating friends, peers, rivals
Waiting for local participants, outsiders, decisions, funds	Satisfaction with home, employment situation, lifestyle

Source: Adapted from Jackson 2000.

Being realistic about these costs and weighing them against the potential benefits (Table 13–1) can help all stakeholders be realistic in their expectations. In the case of Pokorny and his colleagues (Chapter 2), the scientists were the main beneficiaries of monitoring efforts. Yet most of the unpaid costs were borne by local community members. In Bolivia (Chapter 5), on the other hand, the ACM initiative, including the collaborative monitoring, clearly benefited the local people. The ACM team had understood that this was critically important and therefore built the monitoring process around timber sales and showed how it could provide concrete benefits.

The considerable transaction costs for community members may not be the only cause of low participation in collaborative monitoring—concerns about the ACM work itself may cause people to consider the monitoring wasted effort. Santos and her colleagues (Chapter 3) started with 13 families, 3 of which left the program after one year, unconvinced that it benefited them. Conversely, if monitoring shows longer-term benefits, participation in ACM may increase. In the Nepal example (Chapter 8), Ms. B was immediately rewarded for holding the committee accountable to its principles. Making the benefits of the monitoring process public is critical to ACM initiatives. Monitoring can play an important role in sustaining interest in ACM by showing the extent of its impact on forests and people's well-being.

Keeping an Eye on Capacity and Change

Collaborative monitoring engages stakeholders in a range of unfamiliar activities whose initial benefits may not be apparent. Learning how to construct a constitution for forest committees, monitoring the monitoring process, and learning to work with other stakeholders for the first time—all these activities require serious investment in capacity building if collaborative monitoring is to be effective and sustained. It is important to identify who needs which capacities, create a strategy for sustaining the capacity-building process, and consciously phase out external facilitation.

Identifying Necessary Capacities

In ACM, multiple capacities are needed for effective collaborative monitoring. Most obvious is methodological capacity and understanding of monitoring mechanisms that go beyond policing. Monitoring-as-learning, critical reflection, and the cyclical nature of reflect-act (Figure 3, Chapter 1) are particularly important topics that require specific attitudes and skills.

But these are certainly not the only capacities needed. Many natural resource management initiatives have a fairly project-centric perception of participation. ACM-focused initiatives, notwithstanding their collaborative intentions, are no exception. Paudel and Ojha (Chapter 4) describe the problems that occurred when a project-centric and indicator-focused perspective guided the development of a monitoring process, initially blinding them to other institutionalized practices that prohibited learning. Their methodological understanding of indicators could not help them deal with poor representation of stakeholders in decision-making bodies and limited understanding of the rights and responsibilities of devolved forest management. But when they recognized the political and analytical dimensions of monitoring, they were able to reorient their work effectively.

Such "sociopolitical" capacities, though rarely acknowledged as significant for monitoring, are required. They are needed not just by the facilitator—they make it possible for diverse and even opposed stakeholders to conduct analysis, and they enable marginalized groups to speak in decision-making forums. Kusumanto (Chapter 12) describes how people's initial hesitance to learn about political processes shifted once they started understanding those processes and how they could influence them. This made it possible for them to ask more challenging questions about transparency and accountability that they first did not dare to consider.

Building capacity requires identifying which stakeholder group requires what abilities and at what level. Capacity may be needed at the community level or at district or even national levels—depending on where the decision-making forums are and where the analysis of monitoring data takes place. And it may be the facilitators themselves whose capacity needs building. Many of the contributors reflect on their own learning, pointing out assumptions that were challenged as the processes took shape. For example, Paudel and Ojha (Chapter 4) saw the limitations of their focus on indicators, and Pokorny and his colleagues (Chapter 2) realized the rigidity of the criteria and indicators framework.

Ensuring Continuity and Focus

External facilitators often have methodological capacities for monitoring. But local actors may well have a better sense of how to transform information into socially acceptable action. Both types of skills are necessary and may need enhancing. Social and political insights can be gained through context studies (Chapters 3 and 11), for example, and the methodological capacities of local organizations and communities can be developed in training programs. Investment in capacity-building efforts to maintain effective monitoring should focus on the actors who will carry the process in the long term.

The existing capacities, whether methodological or sociopolitical, determine what is possible to achieve in the short term. Monitoring capacities take time to build, and expecting too much too soon inevitably leads to frustrations and depresses motivation. The work in Bolivia (Chapter 5) offers a clear example of an effective strategy, starting slow and small: a simple plan to pay wages to project workers was adapted, without facilitators' assistance, to manage a new project. Several authors note that once a focused monitoring process had become established, spontaneous use of the idea occurred elsewhere (Chapters 4, 6, 7, and 10).

How capacities are built depends on the starting level, the resources available, and the type of capacity in question. Many experiences in this volume have built capacities on the job, using concrete experiences to generate understanding. In some cases, certain individuals or groups received additional training to undertake specific tasks, such as indicator identification or data analysis.

Revising and Modifying Continually

"The wise adapt themselves to circumstances," goes a Chinese proverb, "as water molds itself to the pitcher." ACM stands for adaptive collaborative *management* but could just as well signify adaptive collaborative *monitoring*. Monitoring systems need continual change. As monitoring mechanisms are implemented and capacities are developed, the difference between what is desirable and what is feasible becomes clear and adjustments can be made.

Initial agreements about stakeholders' roles, types of information, and monitoring methods and timing all require regular reviewing to ensure they are still relevant and achievable. Santos and her colleagues (Chapter 3) in the eastern Amazon started with many stakeholders, but after finding the group too unwieldy, they worked with just the rubber tappers. Now it might be possible to open their monitoring process to include other significant stakeholders. Mutimukuru and her colleagues (Chapter 11) give another example, illustrating how terms of reference drafted by stakeholders' representatives were then finalized in a larger gathering. These terms of reference will probably need revision now that implementation is underway.

The design of the monitoring process should allow for periodic review of the methodology. How often will this monitoring review happen and with whom? What questions will be asked? Box 13-1 offers some ideas.

Box 13-1. Checking the Monitoring Process

In Brazil, a three-year process for monitoring joint work in agriculture was reviewed every six months. In one of the case study sites, the team used four criteria to assess the methods and two to assess the indicators, as had been agreed in plenary sessions. If a method or an indicator no longer met the criteria, it was replaced or eliminated. During the six-month reviews, methods were adjusted and tips were exchanged about how to implement certain aspects of the monitoring process, and the importance of ensuring that the information fed into learning was stressed.

Method-related criteria:
- the level of participation of farmers in the collection, collation, and analysis of data and dissemination of findings;
- time demand (for collection, collation, analysis, and dissemination);
- the degree of difficulty of applying the method (mainly for collection and analysis); and
- the potential for others outside the current monitoring group to use the methods (for scaling up and sustainability).

Indicator-related criteria:
- reliability of the information; and
- relevance of the final information (for different audiences: farmers, farmers' union, NGOs, funding agencies, public agencies).

Source: From Sidersky and Guijt 2000.

Phasing Out Facilitation

For learning to be sustained, it must be institutionalized, made the new norm, and embedded locally, and that requires finding a long-term home for the monitoring process. Oyono and his colleagues (Chapter 9) were conscious of the need to phase out their role as external facilitators and found a local organization that worked alongside them and was able to take over once funding for the ACM team ended. The Malawi and Zimbabwe experiences (Chapters 10 and 11) were collaborations with established local actors, in both cases the forest department officials but also local resource management committees; the ACM field teams worked closely with them to develop the monitoring systems, in ways appropriate to the two contexts.

As questioners, mediators, and technical advisers, external facilitators may initially play an important role. But an awareness of the need to phase themselves out and invest in local capacities and relationships can enhance the likelihood that a monitoring process will be sustainable.

Finding Opportunities for Collaborative Monitoring

Collaborative monitoring within the context of ACM initiatives offers many potential benefits, but these benefits are not guaranteed—they are the result of considerable work for the groups involved. The contributors suggest that collaborative monitoring can succeed under the following conducive starting conditions:

- a focused resource issue or problem around which to start;
- sufficient time to develop and implement an agreed process;
- skilled and dedicated facilitation;
- limited diversity of stakeholders (and perspectives) at the onset;
- open discussions about problems in stakeholder interactions;
- innovative solutions to organizing ACM and collaborative monitoring; and
- use of monitoring as a process of analysis and communication, not simply data collection.

The experiences suggest that less is more—at least at first. In most contexts where ACM would be useful, the issues, relationships, and information needs are sufficiently complex and dynamic that a modest start appears wise. Too many issues to monitor may initially overwhelm those responsible for implementation. Too many stakeholders and their diverse interests at the onset can have a stagnating effect. Too many monitoring purposes at once can cause confusion about priorities in data collection, information analysis, communication of findings, and stakeholders' roles.

This collection of hands-on experiences suggests how collaborative monitoring can contribute to adaptive and collaborative forest management under different contexts and conditions. It moves the discussion of adaptive natural resource management forward on two fronts.

First, it clearly shows that collaborative monitoring means more than a technocratic manipulation of indicators, objective measurements, and data hierarchies. A mechanical approach to monitoring that focuses on data collection—as described in much of the resource monitoring literature—does not lay the basis for collective learning. Monitoring means involving diverse stakeholders and building in critical reflection. It is a social and political process that requires negotiation around information identification and analysis.

Second, by showing the great diversity of practice with varying degrees of success and limitations, the authors invite us to revisit the widely held view that collaborative monitoring is a stepwise exercise in data gathering and compilation. Such diversity cannot be standardized with a fixed set of concise steps or a routine, mechanical application of more or less participatory tools. To be effective, adaptive and collaborative monitoring will require political astuteness and a deep understanding of social learning as a collective sense-making process.

The dynamics of the political process, the social and historical patterns of communication and domination, and the nature of the forestry issues determine the extent to which alternative forest management processes can be adaptive and collaborative. Careful design can help offset these unknowns—but adaptation en route will remain essential, as it is for monitoring mechanisms and institutions.

This should not be interpreted as a call for abandoning collaborative monitoring. On the contrary, natural resource management professionals need a grounded understanding of options, outcomes, and challenges to make the most of what collaborative monitoring can offer. The experiences in this volume offer such insights. In so doing, they place monitoring solidly within the realm of social learning, inspiring and guiding those embarking on or immersed in processes of social change.

References

Abbot, J., and I. Guijt. 1998. Changing Views on Change: Participatory Approaches to Monitoring the Environment. SARL Programme Discussion Paper 2. London: IIED.

Assembe, S., and P.R. Oyono. 2004. An Assessment of Social Negotiation as a Tool of Local Management: A Case Study of the Dimako Council Forest (Cameroon). *Scandinavian Journal of Forest Research* 19(4): 78–84.

Baron, N. 1998. *Keeping Watch: Experiences from the Field in Community-Based Monitoring*. A Biodiversity Support Program. Lessons from the Field. Washington, DC: Biodiversity Support Program.

Bennett, L. 1983. *Dangerous Wives and Sacred Sisters: Social and Symbolic Roles of High Caste Women in Nepal*. New York: Columbia University Press.

Biesbrouck, K. 2002. New Perspectives on Forest Dynamics and the Myth of "Communities": Reconsidering Co-management of Tropical Rainforests in Cameroon. *IDS Bulletin* 33(1): 55–64.

Bigombé Logo, P. 2003. The Decentralized Forestry Taxation System in Cameroon: Local Management and State Logic. Working Paper 10 (Environmental Governance in Africa Series). Washington, DC: World Resources Institute.

Blaikie, P.M., J. Cameron, and D. Seddon. 1980. *Nepal in Crisis: Growth and Stagnation at the Periphery*. Oxford: Clarendon Press.

Booth, D., J. Holland, H, Hentschel, P.Lanjouw, and A. Herbert. 1998. Participation and Combined Methods in African Poverty Assessment: Renewing the Agenda. London: Department for International Development.

Borrini-Feyerabend, G., M. Faghi Farvar, J.C. Nguinguiri, and V. Ndangang. 2000. *Co-management of Natural Resources. Organising, Negotiating and Learning-by-Doing*. Eshborn, Germany: GTZ.

Bosch, O.J.H., W.J. Allen, and R.S. Gibson. 1996. Monitoring as an Integral Part of Management and Policy Making. *Proceedings of Symposium "Resource Management: Issues, Visions, Practice."* Lincoln University, New Zealand, 5–8 July 1996, 12–21. Manaaki Whenua, Landcare Research.

Buck, L., E. Wollenberg, and D. Edmunds. 2001. Social Learning in the Collaborative Management of Community Forests: Lessons from the Field. In E. Wollenberg, D. Edmunds, L. Buck, J. Fox, and S. Brodt (eds.), *Social Learning in Community Forests*. Bogor, Indonesia: CIFOR, 1–20.

Buck, L.E., C.C. Geisler, J.W. Schelhas, and E. Wollenberg. 2001. *Biological Diversity: Balancing Interests through Adaptive Collaborative Management*. Boca Raton, FL: CRC Press.

Buhler, W., S. Morse, E. Arthur, S. Bolton, and J. Mann. 2002. *Science, Agriculture and Research. A Compromised Participation?* London: Earthscan.

Carret, J.P. 2000. La réforme de la fiscalité forestière au Cameroun. Débat politique et analyse économique. *Bois et Forêts des Tropiques* 264(2): 37–53.

Center for International Forestry Research (CIFOR). 1999a. The CIFOR Criteria and Indicators Generic Template. The Criteria and Indicators Toolbox Series 2. Bogor, Indonesia: CIFOR.

———. 1999b. The Grab Bag: Supplementary Methods for Assessing Human Well-Being. The Criteria and Indicators Toolbox Series 6. Bogor, Indonesia: CIFOR.

————. 1999c. C&I Toolbox. Bogor, Indonesia: CIFOR.

Chambers, R., and I. Guijt. 1995. PRA: Five Years Later. Where Are We Now? *FTP Newsletter,* 26–27 April 1995.

Colfer, C.J.P. 2005a. *The Complex Forest: Communities, Uncertainty, and Adaptive Collaborative Management.* Washington, DC: Resources for the Future, and Bogor, Indonesia: CIFOR.

————. (ed.). 2005b. *The Equitable Forest: Diversity, Community and Resource Management.* Washington, DC: Resources for the Future, and Bogor, Indonesia: CIFOR.

Colfer, C.J.P., and Y. Byron. 2001. *People Managing Forests: The Link between Human Well-Being and Sustainability.* Washington, DC: Resources for the Future.

Colfer, C.J.P., and R. Prabhu. 2003. Community Forest Management and the Struggle for Equity. Paper presented at the XII World Forestry Congress. September. Quebec, Canada.

Colfer, C.J.P., et al. 1999a. The BAG (Basic Assessment Guide for Human Well-Being). Criteria & Indicators Toolbox Series No. 5. Bogor, Indonesia: CIFOR.

Colfer, C.P.J., R. Prabhu, M. Günter, C. McDougall, N.M. Porro, and R. Porro. 1999b. Who Counts Most? Assessing Human Well-Being in Sustainable Forest Management. Criteria & Indicators Toolbox Series No. 8. Bogor, Indonesia: CIFOR.

Community Forestry Division. 2003. FUG Database. Community Forestry Division, Department of Forests, Kathmandu, Nepal.

Cornwall, A. 2002. Making Spaces, Changing Places: Situating Participation in Development. IDS Working Paper 170. Brighton, UK: IDS.

Davies, R.J. 1998. An Evolutionary Approach to Organisational Learning: An Experiment by an NGO in Bangladesh. In D. Mosse, J. Farrington, and A. Rew (eds.), *Development as Process: Concepts and Methods for Working with Complexity.* London: Routledge.

De Oliveira, N.B. 1999. Community Participation in Developing and Applying Criteria and Indicators of Sustainable and Equitable Forest Management. CIFOR Project Report on Testing Criteria and Indicators for the Sustainable Management of Forests. Bogor, Indonesia: CIFOR.

Defoer, T., S. Kante, et al. 1998. A Participatory Action Research Process to Improve Soil Fertility Management. In H.P. Blume, H. Eger, E. Fleischhauer, A. Hebel, C. Reij, and K.G. Steiner (eds.), *Towards Sustainable Land Use—Furthering Cooperation between People and Institutions.* Proceedings of Ninth International Soil Conservation Organisation Conference. Bonn, Germany: Catena Verlag, 1083–92.

Department of Environment and Natural Resources (DENR). 1996. DENR Administrative Order No. 96–29. Rules and Regulations for the Implementation of Executive Order No. 263, otherwise known as the Community Based Forest Management Strategy (CBFMS). Manila, the Philippines: DENR.

Diaw, C. 2002. L'altérité des tenures forestières: Les théories scientifiques et la gestion des biens communs. Informations et Commentaires. No. 121, octobre-décembre.

Diaw, C., and T. Kusumanto. 2003. Scientists in Social Encounters: The Case of an Engaged Practice of Science. In C. Colfer (ed.), *The Equitable Forest: Diversity and Community in Sustainable Resource Management.* Washington, DC: Resources for the Future, 72–109.

Diaw, C., and P.R. Oyono. 2001. Developing Collaborative Monitoring for Adaptive Co-management of Tropical Forests. Cameroon Extract. Yaoundé, Cameroon: CIFOR.

Diaw, C., F. Tchala Abina, and P.R. Oyono (eds.). Forthcoming. *Adaptive Collaborative Management of Forests and Decentralization Policies in Cameroon.* Bogor, Indonesia: CIFOR.

Dick, B. 1999. What Is Action Research? Occasional Pieces in Action Research Methodology, 2. http://www.scu.edu.au/schools/gcm/ar/arm/op002.html (accessed September 3, 2006).

Dovers, S.R., and C.D. Mobbs. 1997. An Alluring Prospect? Ecology and the Requirements of Adaptive Management. In N. Klomp and I. Lunt (eds.), *Frontiers in Ecology: Building the Links.* Oxford, UK: Elsevier Science, 39–52.

Easterby-Smith, M.P.V., and M. Lyles (eds.). 2003. *The Blackwell Handbook of Organizational Learning and Knowledge Management.* Oxford, UK: Blackwell.

Ebene, Y., and P.R. Oyono. 2000. Rapport-synthèse sur l'atelier de lancement du Programme ACM au Cameroun. Yaoundé, Cameroon: CIFOR.

Edmunds, D., and E. Wollenberg. 2001. A Strategic Approach to Multistakeholder Negotiations. *Development and Change* 32(2): 231–53.

Edwards, J. 2001. *The Things We Steal from Children.* Paper presented at Ninth International Conference on Thinking. January. Auckland, New Zealand.

Edwards, M. 1997. *Organizational Learning in Non-Governmental Organizations: What Have We Learned?* Washington, DC: World Bank.

Efoua, S. 2002. Etat des lieux du site de recherche de Lomié. Yaoundé, Cameroon: CIFOR.

Engel, P.G.H., A. Hoeberichts, and L. Umans. 2001. Accommodating Multiple Interests in Local Forest Management: A Focus on Facilitation, Actors and Practices. *International Journal of Agricultural Resources, Governance and Ecology* 13(4): 306–26.

Estrella, M., and J. Gaventa. 1997. *Who Counts Reality? Participatory Monitoring and Evaluation: A Literature Review.* Discussion paper prepared for International Workshop on Participatory Monitoring and Evaluation, Silang (the Philippines), November.

Estrella, M., et al. 2000. *Learning from Change: Issues and Experiences in Participatory Monitoring and Evaluation.* London, UK: Intermediate Technology Publications.

Etoungou, P. 2003. Decentralization Viewed from Inside: The Implementation of Community Forests in East Cameroon. Working Paper 12. Environmental Governance in Africa Series. Washington, DC: World Resources Institute.

Fisher, R.J., R. Prabhu, and C. McDougall (eds.). Forthcoming. *Adaptive Collaborative Management of Community Forests in Asia: Experiences from Nepal, Indonesia and the Philippines.* Bogor, Indonesia: CIFOR.

Fomété, T. 2001. The Forestry Taxation System and the Involvement of Local Communities in Forest Management in Cameroon. RDFN Network Paper 25b. London: Overseas Development Institute, 17–27.

Forest Users Group Forest Management Project. 2000. Supply and Demand Relationships in Community Forests. FFMP Discussion Paper 4. University of Reading, UK: IRDD.

Gaventa, J. 2006. Reflections on the Uses of the "Power Cube" Approach for Analyzing the Spaces, Places and Dynamics of Civil Society Participation and Engagement. Prepared for Dutch CFA evaluation "Assessing Civil Society Participation as supported by Cordaid, Hivos, Novib and Plan Netherlands" and for Power, Participation and Change Programme Participation Group, Institute of Development Studies.

Ghimire, K.B., and M.P. Pimbert. 1997. Social Change and Conservation: An Overview of Issues and Concepts. In K.B. Ghimire and M.P. Pimbert (eds.), *Social Change and Conservation.* London: Earthscan, 1–45.

Global Forest Watch. n.d. Comprehensive Maps Provide Key Tools to Manage Northern Forest Frontier. http://www.globalforestwatch.org/english/index.htm.

Groot, A., and M. Maarleveld. 2000. Demystifying Facilitation of Multi-Actor Learning Processes. Gatekeeper Series 89. London: IIED.

Guijt, I. Forthcoming. *Seeking Surprise: The Role of Monitoring to Trigger Learning in Collective Rural Resource Management.* Ph.D. thesis. Wageningen, Netherlands: Wageningen University and Research Centre.

Gunderson, L., C.S. Holling, and S.S. Light. 1995. *Barriers and Bridges to the Renewal of Ecosystems and Institutions.* New York: Columbia University Press.

Haggith, M., and R. Prabhu. 2003. Unlocking Complexity: The Importance of Idealization in Simulation Modelling. *Small-Scale Forest Economics, Management and Policy* 2(2): 293–312.

Hagmann, J. 1999. *Learning Together for Change: Facilitating Innovation in Natural Resource Management through Learning Process Approaches in Rural Livelihoods in Zimbabwe.* Weikersheim, Germany: Margraf Verlag.

Hartanto, H., M.C.B. Lorenzo, and A.L. Frio. 2002. Collective Action and Learning in Developing a Local Monitoring System. *International Forestry Review* 4(3): 184–95.

Herweg, K., and K. Steiner. 2002. *Impact Monitoring and Assessment. Instruments for Use in Rural Development Projects with a Focus on Sustainable Land Management,* vols. 1, 2. Berne: CDE and Bonn: GTZ.

Hinchcliffe, F., J. Thompson, J. Pretty, I. Guijt, and P. Shah. 1999. *Fertile Ground: The Impact of Participatory Watershed Development.* London: Intermediate Technology Publications.

International Association for Impact Assessment (IAIA). 2003. Social Impact Assessment. International Principles. http://www.iaia.org/Members/Publications/Guidelines_Principles/SP2.pdf (accessed 30 August 2006).

International Tropical Timber Organization (ITTO). 1992. Criteria and Indicators for Sustainable Management of Natural Tropical Forests. ITTO Policy Development Series 3. Yokohama: ITTO.

International Union for Conservation of Nature and Natural Resources (IUCN). 1997. *Un enfoque para la evaluación del progreso hacia la sostenibilidad. Serie Herramientas y Capacitación. Para instituciones, equipos de campo y agencies colaboradoras.* Gland: World Conservation Union.

Jackson, E.T. 2000. The Front-End Costs and Downstream Benefits of Participatory Evaluation. In O. Feinstein and R. Picciotto (eds.), *Evaluation and Poverty Reduction.* Proceedings from a World Bank Conference. Washington, DC: World Bank, 115–26.

Jiggins, J., and N. Roling 2000. Adaptive Management: Potential and Limitations for Ecological Governance. *International Journal of Agricultural Resources, Governance and Ecology* 1(1): 28–42.

Kanel K.R., and B.K. Pokharel. 2002. Community Forestry in Nepal: The Potential Contribution of Adaptive and Collaborative Approaches to Management. Paper presented at FAO RAP Workshop on Adaptive Collaborative Management of Community Forests: An Option for Asia. September. Bangkok, Thailand.

Kemmis, S., and R. McTaggart. 1988. *The Action Research Planner.* Third edition. Geelong, Australia: Deakin University.

Khadka, C., M. Gurung, and N. Tumbahangphe. 2003. Addressing Equity Issues in Community Forestry. Research report. Kathmandu, Nepal: ICIMOD.

Klein, M., B. Salla, and J. Kok. 2001. Attempts to Establish Community Forests in Lomié, Cameroon. Rural Development Network Paper 25b. London: Overseas Development Institute.

Klouda, T. 2004. Thinking Critically, Speaking Critically. Unpublished paper. http://www.tonyklouda.pwp.blueyonder.co.uk/Critical%20thinking%20TK%2018%20Nov%2004.pdf (accessed 3 September 2006).

Kolb, D.A. 1984. *Experiential Learning: Experience as the Source of Learning and Development.* Englewood Cliffs, NJ: Prentice-Hall.

Kusumanto, T., S. Hakim, L. Yuliani, Y. Indriatmoko, and H. Adnan. 2002a. Final Report to the Asian Development Bank: Indonesia Country Report: Planning for Sustainability of Forests through Adaptive Co-Management. Bogor, Indonesia: CIFOR.

———. 2002b. Summary: Adaptive Collaborative Management Research in Indonesia: Expectation, Outcomes, Lessons and Implications. Paper presented at FAO RAP Workshop on Adaptive Collaborative Management of Community Forests: An Option for Asia. September. Bangkok, Thailand.

Kusumanto, T., E.L. Yuliani, P. Macoun, Y. Indriatmoko, and H. Adnan. 2005. Learning to Adapt: Managing Forests Together in Indonesia. Bogor, Indonesia: CIFOR, YGB, and PSHK-ODA.

Lammerts van Bueren, E.M., and E.M. Blom. 1997. Hierarchical Framework for Formulation of Sustainable Forest Management Standards: Principles, Criteria, Indicators. Strategy (theme) paper 3. Wageningen, Netherlands: Tropenbos Foundation.

Leach, M. I. Scoones, and B. Wynne (eds.). 2005. *Science and Citizens: Globalization and the Challenge of Engagement.* London: Zed Books.

Lee, K. 1993. *Compass and Gyroscope: Integrating Science and Politics for the Environment.* Washington, DC: Island Press.

Lee, K.N. 1999. Appraising Adaptive Management. *Conservation Ecology* 3(2): 3–13.

Leeuwis, C., and R. Pyburn (eds.). 2002. *Wheelbarrows Full of Frogs. Social Learning in Rural Resource Management.* Assen, Netherlands: Royal van Gorcum.

Li, T. 2002. Engaging Simplifications: Community-Based Resource Management, Market Processes and State Agendas in Upland Southeast Asia. *World Development* 30(2): 265–83.

Lindenberg, M., and C. Bryant. 2001. *Going Global: Transforming Relief and Development NGOs.* Bloomfield, CT: Kumarian Press.

Lorenzo, M.C. 2001. Historical Trends of ACM Site in Palawan. ACM Project Internal Research Report. Bogor, Indonesia: CIFOR.

Maarleveld, M. 2003. Social Environmental Learning for Sustainable Natural Resource Management. Ph.D. thesis. Wageningen, Netherlands: Wageningen University and Research Centre.

Maarleveld, M., and C. Dangbégnon. 1999. Managing Natural Resources in the Face of Evolving Conditions: A Social Learning Perspective. *Agriculture and Human Values* 16(3): 267–80.

MacGillivray, A., C. Weston, and C. Unsworth. 1998. *Communities Count! A Step-by-Step Guide to Community Sustainability Indicators.* London: New Economics Foundation.

Mala, A. W. 2002. Rapport intermédiaire sur le statut et les étapes accomplies: monitoring standard and indicators. Yaoundé, Cameroon: CIFOR.

Mala, A. W., J. Nguiébouri, and G. Mato. 2003. Rapport d'atelier de formation sur l'usage des outils d'aide à la decision (Co-View et Co-Learn). Yaoundé, Cameroon: CIFOR.

Malla, Y. B. 2000. Impact of Community Forestry Policy on Rural Livelihoods and Food Security in Nepal. *Unasylva* 202: 37–45.

———. 2001. Changing Policies and the Persistence of Patron-Client Relations in Nepal: Stakeholders' Response to Changes in Forest Policies. *Environmental History* 6(2): 287–307.

Malla, Y. B., R. Barnes, K. Paudel, A. Lawrence, H. Ojha, and K. Green. 2002. Common Property Forest Resource Management in Nepal: Developing Monitoring Systems for Use at the Local Level. Project Report. University of Reading, UK: IRDD with ForestAction, Nepal, and ECI, Oxford.

Matose, F., and R. Prabhu (eds.). Forthcoming. *Adaptive Resource Management in Zimbabwe: What Prospects for Policy Change?* Bogor, Indonesia: CIFOR.

McCarthy, J. F. 2004. Changing to Gray: Decentralization and the Emergence of Volatile Socio-Legal Configurations in Central Kalimantan, Indonesia. *World Development* 32(7): 1199–223.

McDougall, C. 2000. Draft Working Model of ACM (Nr. 2), Local People, Devolution and Adaptive Co-Management Program. February. Bogor, Indonesia: CIFOR.

———. 2003. Improving Livelihoods and Equity in Community Forestry in Nepal: The Role of Adaptive Collaborative Management. Research Proposal to the International Development Research Centre. Bogor, Indonesia: CIFOR.

McDougall, C., and A. Braun. 2003. Navigating Complexity, Diversity and Dynamism: Reflections on Research for Natural Resource Management. In B. Pound, S. Snapp, C. McDougall and A. Braun (eds.), *Managing Natural Resources for Sustainable Livelihoods: Uniting Science and Participation.* London: Earthscan.

McDougall, C., R. Prabhu, and T. Kusumanto. 2003. Participatory Action Research on Adaptive Collaborative Management of Community Forests: A Multi-Country Model. In B. Pound, S. Snapp, C. McDougall, and A. Braun (eds.), *Managing Natural Resources for Sustainable Livelihoods: Uniting Science and Participation.* London: Earthscan, 189–91.

McDougall, C., Kaski ACM Team, New ERA ACM Team, and ForestAction. 2002a. Planning for the Sustainability of Forests through Adaptive Co-Management: Nepal Country Report. ACM Project/MFSC Internal Research Report. Bogor, Indonesia: CIFOR.

McDougall, C., and Nepal, Philippines, and Indonesia ACM Teams. 2002b. Roles, Relations, Access and (In)Equity: Insights into Gender and Diversity in Community Forest Management in Nepal, Indonesia and the Philippines. Internal Research Report. Bogor, Indonesia: CIFOR.

McDougall, C., New Era ACM Team, Kaski ACM Team, and ForestAction ACM Team. 2002c. Adaptive and Collaborative Management Research Project in Nepal: An Overview of the Project and Preliminary Lessons. Research Summary. Paper prepared for the National Workshop on Learning from Community Forestry: How Can Adaptive Collaborative Management Approaches Enhance Livelihoods, Equity and Forests? September 2002. Kathmandu, Nepal.

McGee, R. 2004. Constructing Poverty Trends in Uganda: A Multidisciplinary Perspective. *Development and Change* 35(3): 499–523.

Mendoza, G. A., and P. Macoun, with R. Prabhu, D. Sukadri, H. Purnomo, and H. Hartanto. 1999. Guidelines for Applying Multi-Criteria Analysis to the Assessment of Criteria and Indicators. Criteria and Indicators Toolbox Series No 9. Bogor, Indonesia: CIFOR.

Milbrath, L. W. 1989. *Envisioning a Sustainable Society: Learning Our Way Out.* New York: State University of New York Press.

Milol, A., and J.M. Pierre. 2000. Impact de la fiscalité décentralisée sur le développement local et les pratiques d'utilisation des ressources forestières au Cameroun. Yaoundé, Cameroon: World Bank.

Muhtaman, D.R., C.A. Siregar, and P. Hopmans. 2000. *Criteria and Indicators for Sustainable Plantation Forestry in Indonesia.* Bogor, Indonesia: CIFOR.

Narayan, D., and L. Srinivasan. 1994. Participatory Development Tool Kit: Training Materials for Agencies and Communities. Water and Sanitation Program. Washington, DC: World Bank.

Nemarundwe, N., W. de Jong, and P. Cronkleton. 2003. *Future Scenarios as an Instrument for Forest Management: Manual for Training Facilitators of Future Scenarios.* Bogor, Indonesia: CIFOR.

Nunes W., B. Pokorny, and M. Mattos. 2004. Levantamento das condições de vida por produtores familiares em três municípios do Nordeste Paraense utilizando critérios e indicadores de sustentabilidade. Final report. Bélem, Brazil: EMBRAPA.

Ojha, H., B. Pokharel, and K. Paudel with C. McDougall. 2002. Stakeholder Collaboration, Adaptive Management and Social Learning: A Comparative Review of Eight Community Forestry Sites in Nepal. Research report. Kathmandu, Nepal: ForestAction and CIFOR.

Ottke, C., P. Kristensen, D. Maddox, and E. Rodenburg. 2000. *Monitoring for Impact: Lessons on Natural Resources Monitoring from 13 NGOs: Volume 1.* Washington, DC: World Resources Institute.

Oyono, P.R., G. Akwah, M.C. Diaw, C. Jum, A.M. Tiani, W. Mala, and S. Assembe. 2003a. Developing Collaborative Monitoring for Adaptive Co-Management of Tropical Forests (ACM). Final report, Cameroon. Yaoundé, Cameroon: CIFOR.

Oyono, P.R., W. Mala, and J. Tonyé. 2003b. Adaptation versus Rigidity: Contribution to the Debate on Agricultural Viability and Forest Sustainability in Southern Cameroon. *Culture & Agriculture* 25(2).

Paudel, K., and Ojha, H. 2002. A Review of Monitoring Systems and Practices in Community Forestry at the Local Level. Final report. Kathmandu, Nepal: ForestAction, and Bogor, Indonesia: CIFOR.

Paudel, K., H. Naupane, H. Ojha, and R. Barnes. 2003. Action and Learning Processes for Common Property Forest Management: An Assessment of the Current Status and Impacts of Local Action Learning Processes for Common Property Forest Management Developed through a Participatory Action Research Project in Nepal. Followup study for NRSP Project PD119: Post R7514. Kathamandu, Nepal: ForestAction. http://www.forestaction.org/publications/2_Dev%27t%20Action%20and%20Learning%20Reports/2_6.pdf. (accessed 3 September 2006).

Pettit, J., and L. Roper. 2003. Development and the Learning Organisation: An Introduction. In J. Pettit, L. Roper and D. Eade (eds.), *Development and the Learning Organisation* (Development in Practice Readers Series). Oxford, UK: Oxfam Publications.

Pokharel, B.K., and J. Grosen. 2000. Governance, Monitoring and Evaluation, Joint Technical Review of Community Forestry in Nepal. Issue Paper 5. Kathmandu, Nepal: Ministry of Forest and Soil Conservation.

Pokorny, B., G. Cayres, W. Nunes, D. Segebart, R. Drude, and M. Steinbrenner. 2003a. *Adaptive Collaborative Management: Criteria and Indicators to Assess Sustainability. Manejo Colaborativo Adaptativo: Criteria e indicadores para avaliar sustentabilidade.* Bogor, Indonesia: CIFOR.

Pokorny, B., G. Cayres, W. Nunes, D. Segebart, and R. Drude. 2003b. First Experiences with Adaptive Co-Management in Pará, Brazilian Amazon. In C. Sabogal and N. Silva (eds.), *Integrated Management of Neotropical Rain Forests by Industries and Communities.* Belém. Brazil: EMBRAPA, 258–80.

Pokorny, B., R. Prabhu, C. McDougall, and R. Bauch. 2004. Local Stakeholders' Participation in Developing Criteria and Indicators for Sustainable Forest Management. *Journal of Forestry* 102(1): 25–40.

Poulsen, J., G. Applegate, and D. Raymond. 2001. Linking C&I to a Code of Practice for Industrial Tropical Tree Plantations. Bogor, Indonesia: CIFOR.

Prabhu, R. 2002. Developing Collaborative Monitoring for Adaptive Co-Management of Tropical African Forests. Contract B7–6201/99–05/FOR. Interim report to the European Union for period January 1–December 31, 2001. Harare, Zimbabwe: CIFOR.

————. 2003. Developing Collaborative Monitoring for Adaptive Co-Management of Tropical African Forests. Contract B7–6291/99–05/FOR. Final technical report for period January 1, 2000–December 31, 2002. Bogor, Indonesia: CIFOR.

Prabhu, R., C.J.P. Colfer, P. Venkateswarlu, L.C. Tan, R. Soekmadi, and E. Wollenberg. 1996. Testing Criteria and Indicators for the Sustainable Management of Forests: Phase I Final Report. Bogor, Indonesia: CIFOR.

Prabhu, R., H.J. Ruitenbeek, T.J.B. Boyle, and C.J.P. Colfer. 1998a. *Between Voodoo Science and Adaptive Management: The Role and Research Needs for Indicators of Sustainable Forest Management.* Paper presented at IUFRO Conference on Indicators for Sustainable Forest Management. August. Melbourne, Australia.

Prabhu, R., C.J.P. Colfer, and G. Shepherd. 1998b. Criteria and Indicators for Sustainable Forest Management: New Findings from CIFOR's Forest Management Unit Level Research. Rural Development Forestry Network Paper No. 23. London: Overseas Development Institute Rural Development Forestry Network.

Prabhu, R., C.J.P. Colfer, and R.G. Dudley. 1999. Guidelines for Developing, Testing and Selecting Criteria and Indicators for Sustainable Forest Management: A C&I Developer's Reference. Criteria and Indicators Toolbox Series No.1. Bogor, Indonesia: CIFOR.

Prabhu, R., F. Matose, L. Mwabumba, J. Kamoto, H. Madevu, J. Milner, R. Nyirenda, and W. Standa-Gunda. 2002. Developing Indicator-Based Collaborative Monitoring Arrangements to Promote Adaptive Community Based Forest Management. Final technical report to DFID. Harare, Zimbabwe: CIFOR.

Probst, K. 2002. Participatory Monitoring and Evaluation: A Promising Concept in Participatory Research? Lessons from Two Case Studies in Honduras. Weikersheim, Germany: Margraf Publishers.

Purnomo, H., et al. 2000. CIMAT Criteria and Indicators Modification and Adaptation Tool–Version 2. Criteria and Indicators Toolbox Series 3. CD-ROM and user manual. Bogor, Indonesia: CIFOR.

Ritchie, B., C. McDougall, M. Haggith, and N. Burford de Oliveira. 2000. *Criteria and Indicators for Community Managed Forest Landscapes: An Introductory Guide.* Bogor, Indonesia: CIFOR.

Robiglio, V., A.W. Mala, and M.C. Diaw. 2003. Mapping Landscapes: Integrating GIS and Social Science Methods to Model Human-Nature Relationships in Southern Cameroon. *Small-Scale Forest Economics, Management and Policy Journal* 2(2): 171–84.

Roche, C. 1999. *Impact Assessment for Development Agencies: Learning to Value Change.* Oxford: Oxfam Publishing.

Rocheleau, D. 2003. Participation in Context: What's Past, What's Present and What's Next. In B. Pound, S. Snapp, C. McDougall, and A. Braun (eds.), *Managing Natural Resources for Sustainable Livelihoods: Uniting Science and Participation.* London: Earthscan.

Roe, E., M. v. Eeten, and P. Gratzinger. 1999. Threshold-Based Resource Management: The Framework, Case Study and Application, and Their Implications. Report to the Rockefeller Foundation. Berkeley: University of California.

Ruitenbeek, J., and C. Cartier. 2001. The Invisible Wand: Adaptive Co-management as an Emergent Strategy in Complex Bio-economic Systems. CIFOR Occasional Paper 34. October.

Sankar, S., P.C. Anil, and M. Amruth. 2000. *Criteria and Indicators for Sustainable Plantation Forestry in India.* Bogor, Indonesia: CIFOR.

Santos, M.C. Dos, N. Porro, and M.P. Schmink. 2001. Relatório do Co-manejo Adaptativo no Acre—Fase I. Report written for International Center for Forestry Research. Bogor, Indonesia: CIFOR.

Santos, M.C. Dos, N. Porro, S. Stone, and M. Schmink. 2003. Os Desafios do Co-Manejo Colaborativo e Adaptativo de Madeira na Amazônia Brasileira: Estudo de Caso do Projeto Agroextrativista Porto Dias, Acre. Discussion document (unpublished).

Sarin, M., with N.M. Singh, N. Sundar, and R.K. Bhogal. 2003. *Devolution as a Threat to Democratic Decision-Making in Forestry? Findings from Three States in India.* Working Paper 197. London: Overseas Development Institute.

Selener. D. 1997. *Participatory Action Research and Social Change,* 2nd ed. New York: Cornell University Press.

Sidersky, P., and I. Guijt. 2000. Experimenting with Participatory Monitoring in Northeast Brazil: The Case of AS-PTA's Projeto Paraíba. In M. Estrella, et al. (eds.), *Learning from Change: Issues and Experiences in Participatory Monitoring and Evaluation*. London: Intermediate Technology Publications.

Siebert, K., and M. Daudelin. 1999. *The Role of Reflection in Managerial Learning Theory, Research, and Practice*. Westport, CT: Greenwood Press.

Sithole, B. 2002. *Where the Power Lies: Multiple Stakeholder Politics over Natural Resources. A Participatory Methods Guide*. Bogor, Indonesia: CIFOR.

Springate-Baginski, O., J.G. Soussan, O.P. Dev, N.P. Yadav, and E. Kiff. 1999. Community Forestry in Nepal: Impacts on Common Property Resource Management. Environment and Development Series 3. Leeds, UK: School of the Environment, University of Leeds.

Tratado de Cooperación Amazonica (TCA). 2001. Processo de Tarapoto sobre critérios e indicadores de sustentabilidade da floresta Amazônica. Consulta Brasileira de Validação. Brasília, Brazil: Secretaria Pro Tempore/ Brazilian Ministry for Environment.

UNCED. 1992. *Agenda 21*. Adopted June 14, Rio de Janeiro. http://www.un.org/esa/sustdev/documents/agenda21/english/agenda21toc.htm (accessed 9 November 2006).

UNDP (United Nations Development Programme). Undated. Monitoring and Evaluation Framework of the Global Environment Facility Small Grants Programme. New York: UNDP.

Upreti, B. 2001. Beyond Rhetorical Success: Advancing the Potential for the Community Forestry Programme in Nepal to Address Equity Concerns. In E. Wollenberg, D. Edmunds, L. Buck, J. Fox, and S. Brodt (eds.), *Social Learning in Community Forests*. Bogor, Indonesia: CIFOR, 189–207.

Waddell, S. 2005. *Societal Learning and Change. How Governments, Business and Civil Society are Creating Solutions to Complex Multi-stakeholder Problems*. Sheffield, UK: Greenleaf Publishing Ltd.

Wallace, T., and J. Chapman. 2003. Some Realities behind the Rhetoric of Downward Accountability. Working paper presented at INTRAC Fifth Evaluation Conference, Holland. April. http://www.intrac.org/docs/Wallace_Chapman.pdf (accessed 3 September 2006).

Winrock International. 2002. Emerging Issues in Community Forestry in Nepal. Kathmandu, Nepal: Winrock International.

Wollenberg, E. 2000. Methods for Estimating Forest Income and Their Challenges. *Society and Natural Resources* 13(8): 777–95.

Wollenberg, E., D. Edmunds, and L. Buck. 2000. *Anticipating Change: Scenarios as a Tool for Adaptive Forest Management. A Guide*. Bogor, Indonesia: CIFOR.

Wollenberg, E., J. Anderson, and D. Edmunds. 2001a. Pluralism and the Less Powerful: Accommodating Multiple Interests in Local Forest Management. *International Journal of Agriculture, Resources, Governance and Ecology* 1(3/4): 199–222.

Wollenberg, E., D. Edmunds, L. Buck, J. Fox, and S. Brodt (eds.). 2001b. *Social Learning in Community Forests*. Bogor, Indonesia: CIFOR.

Wollenberg, E., B. Campbell, S. Shackleton, and D. Edmunds. 2003. Central Control of Local Resource Management: The Impacts of Devolution. *ETFRN News* 39: 98–100.

Woodhill, J. 2002. Sustainability, Social Learning and the Democratic Imperative. Lessons from the Australian Landcare Movement. In C. Leeuwis and R. Pyburn (eds.), *Wheelbarrows Full of Frogs. Social Learning in Rural Resource Management*. Assen, Netherlands: Royal van Gorcum.

———. Forthcoming. M&E as Learning: Rethinking the Dominant Paradigm. In J.D. Graaff, C. Pieri, S. Sombatpanit, and J. Cameron (eds.), *Monitoring and Evaluation of Soil Conservation and Watershed Development Projects*. World Association of Soil and Water Conservation.

Index